EINSTEIN'S CLOCKS, POINCARÉ'S MAPS

EINSTEIN'S CLOCKS, POINCARÉ'S MAPS

Empires of Time

Peter Galison

W. W. Norton & Company   New York London

 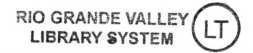

For information about permission to reproduce selections from this book, write to
Permissions, W. W. Norton & Company, Inc.
500 Fifth Avenue, New York, NY 10110

Manufacturing by The Haddon Craftsmen, Inc.
Book design by Chris Welch
Production manager: Anna Oler

Library of Congress Cataloging-in-Publication Data
Galison, Peter Louis.
Einstein's clocks and Poincaré's maps : empires of time / by Peter Galison. — 1st ed.
p. cm.
Includes bibliographical references and index.
ISBN 0-393-02001-0
1. Time. 2. Relativity (Physics) 3. Einstein, Albert, 1879–1955. 4. Poincaré, Henry, 1854–1912. I. Title.

QB209 .G35   2003 2002155114
                                                      529—dc21

W. W. Norton & Company, Inc., 500 Fifth Avenue, New York, N.Y. 10110
www.wwnorton.com

W. W. Norton & Company Ltd.
Castle House, 75/76 Wells Street, London W1T 3QT

1 2 3 4 5 6 7 8 9 0

For Sam and Sarah,
who have taught me the right use of time

CONTENTS

ACKNOWLEDGMENTS

I have benefited enormously from discussions with many students and colleagues. It is a privilege to be able in particular to thank David Bloor, Graham Burnett, Jimena Canales, Debbie Coen, Olivier Darrigol, Lorraine Daston, Arnold Davidson, James Gleick, Michael Gordin, Daniel Goroff, Gerald Holton, Michael Janssen, Bruno Latour, Robert Proctor, Hilary Putnam, Juergen Renn, Simon Schaffer, Marga Vicedo, Scott Walter, and especially Caroline Jones, for their many thoughtful comments. Over the years it has been a pleasure to have learned as well from many discussions with Einstein scholars Martin Klein, Arthur Miller, and John Stachel. Though short in pages, the manuscript and picture preparation were long and certainly could not have been done without the help of research assistants Doug Campbell, Evi Chantz, Robert Macdougall, Susanne Pickert, Sam Lipoff, Katia Scifo, Hanna Shell, and Christine Zutz. Particular thanks go to my Norton editor, Angela von der Lippe, and my agent, Katinka Matson, for good ideas and much encouragement. Amy Johnson and Carol Rose offered many editorial improvements. Finally, I owe a great deal to the many archivists who graciously helped in my research — especially at the Observatoire de Paris, the Archives Nationales, the Archives de la Ville de Paris, the New York Public Library, the U.S. National Archives, the National Archives of Canada, the Bürgerbibliothek Bern, and the Stadtarchiv Bern.

EINSTEIN'S
CLOCKS,
POINCARÉ'S
MAPS

Chapter 1

SYNCHRONY

TRUE TIME WOULD never be revealed by mere clocks—of this Newton was sure. Even a master clockmaker's finest work would offer only pale reflections of the higher, absolute time that belonged not to our human world, but to the "sensorium of God." Tides, planets, moons—everything in the Universe that moved or changed—did so, Newton believed, against the universal background of a single, constantly flowing river of time. In Einstein's electrotechnical world, there was no place for such a "universally audible tick-tock" that we can call time, no way to define time meaningfully except in reference to a definite system of linked clocks. Time flows at different rates for one clock-system in motion with respect to another: two events simultaneous for a clock observer at rest are not simultaneous for one in motion. "Times" replace "time." With that shock, the sure foundation of Newtonian physics cracked; Einstein knew it. Late in life, he interrupted his autobiographical notes to apostrophize Sir Isaac with intense intimacy, as if the intervening centuries had vanished; reflecting on the absolutes of space and time that his theory of relativity had shattered, Einstein wrote: "Newton, forgive me ['Newton, verzeih' mir']; you found the only way which, in your age, was just about possible for a man of highest thought—and creative power."[1]

At the heart of this radical upheaval in the conception of time lay an extraordinary yet easily stated idea that has remained dead-center in physics, philosophy, and technology ever since: *To talk about time, about simultaneity at a distance, you have to synchronize your clocks. And if you want to synchronize two clocks, you have to start with one, flash a signal to the other, and adjust for the time that the flash takes to arrive.* What could be simpler? Yet with this procedural definition of time, the last piece of the relativity puzzle fell into place, changing physics forever.

This book is about that clock-coordinating procedure. Simple as it seems, our subject, the coordination of clocks, is at once lofty abstraction and industrial concreteness. The materialization of simultaneity suffused a turn-of-the-century world very different from ours. It was a world where the highest reaches of theoretical physics stood hard by a fierce modern ambition to lay time-bearing cables over the whole of the planet to choreograph trains and complete maps. It was a world where engineers, philosophers, and physicists rubbed shoulders; where the mayor of New York City discoursed on the conventionality of time, where the Emperor of Brazil waited by the ocean's edge for the telegraphic arrival of European time; and where two of the century's leading scientists, Albert Einstein and Henri Poincaré, put simultaneity at the crossroads of physics, philosophy, and technology.

Einstein's Times

For its enduring echoes, Einstein's 1905 article on special relativity, "On the Electrodynamics of Moving Bodies," became the best-known physics paper of the twentieth century, and his dismantling of absolute time is its crowning feature. Einstein's argument, as usually understood, departs so radically from the older, "practical" world of classical mechanics that the paper has become a model of revolutionary thought, seen as fundamentally detached from a material, intuitive relation to the world. Part philosophy and part physics, Einstein's rethinking of simultaneity has come to stand for the irresolvable break between modern physics and all earlier framings of time and space.

Einstein began his relativity paper with the claim that there was an asymmetry in the then-current interpretation of electrodynamics, an asymmetry not present in the phenomena of nature. Almost all physicists around 1905 accepted the idea that light waves—like water waves or sound waves—must be waves *in* something. In the case of light waves (or the oscillating electric and magnetic fields that constituted light), that something was the all-pervasive *ether*. Most late-nineteenth-century physicists considered the ether to be one of the great ideas of their era, and they hoped that once properly understood, intuited, and mathematized, the ether would lead science to a unified picture of phenomena from heat and light to magnetism and electricity. Yet it was the ether that gave rise to the asymmetry that Einstein rejected.[2]

In physicists' usual interpretation, Einstein wrote, a moving magnet approaching a coil at rest in the ether produces a current indistinguishable from the current generated when a moving coil approaches a magnet at rest in the ether. But the ether itself could not be observed, so in Einstein's view there was but a single observable phenomenon: coil and magnet approach, producing a current in the coil (as evidenced by the lighting of a lamp). But in its then-current interpretation, electrodynamics (the theory that included Maxwell's equations—describing the behavior of electric and magnetic fields—and a force law that predicted how a charged particle would move in these fields) gave two different explanations of what was happening. Everything depended on whether the coil or the magnet was in motion with respect to the ether. If the coil moved and the magnet remained still in the ether, Maxwell's equations indicated that the electricity in the coil experienced a force as the electricity traversed the magnetic field. That force drove the electricity around the coil lighting the lamp. If the magnet moved (and coil stayed still), the explanation changed. As the magnet approached the coil, the magnetic field near the coil grew stronger. This changing magnetic field (according to Maxwell's equations) produced an electric field that drove the electricity around the stationary coil and lit the lamp. So the standard account gave *two* explanations depending on whether one viewed the scene from the point of view of the magnet or the point of view of the coil.

As Einstein reframed the problem there was *one* single phenomenon: coil and magnet approached each other, lighting the lamp. As far as he was concerned, *one* observable phenomenon demanded *one* explanation. Einstein's goal was to produce that single account, one that did not refer to the ether at all, but instead depicted the two frames of reference, one moving with the coil and one with the magnet, as offering no more than two perspectives on the same phenomenon. At stake, according to Einstein, was a founding principle of physics: relativity.

Almost three hundred years earlier, Galileo had similarly questioned frames of reference. Picturing an observer in a closed ship's cabin, borne smoothly across the seas, Galileo reasoned that no mechanical experiment conducted in a below-deck laboratory would reveal the motion of the ship: fish would swim in a bowl just as they would were the bowl back on land; drops would not deviate from their straight drip to the floor. There simply was no way to use any part of mechanics to tell whether a room was "really" at rest or "really" moving. This, Galileo insisted, was a basic feature of the mechanics of falling bodies that he had helped create.

Building on this traditional use of the relativity principle in mechanics, Einstein in his 1905 paper raised relativity to a principle, asserting that physical processes are independent of the uniformly moving frame of reference in which they take place. Einstein wanted the relativity principle to include not only the mechanics of drops dripping, balls bouncing, and springs springing but also the myriad effects of electricity, magnetism, and light.

This relativity postulate ("no way to tell which unaccelerated reference frame was 'truly' at rest") gave rise to an additional assumption that proved even more surprising. Einstein noted that experiments did not show light traveling at any speed other than 300,000 kilometers per second. He then *postulated* that this was

always so. Light, Einstein said, always travels by us at the same measured speed — 300,000 kilometers per second — *no matter how fast
the light source is traveling*. This was certainly not how everyday
objects behaved. A train approaches and the conductor throws a
mailbag forward toward a station; it goes without saying that someone standing on the station platform sees the bag approach at the
speed of the train *plus* the speed at which the conductor habitually
hurls mail. Einstein insisted that light was different: stand, your
lantern raised, at a fixed distance from me and I see the light travel
by me at 300,000 kilometers per second. Hurtle toward me on a
train, even one moving at 150,000 kilometers per second (half the
speed of light), and I still see the light from your lantern go by me at
300,000 kilometers per second. According to Einstein's second postulate, the speed of the source does not matter to the velocity of light.

Both of these postulates would have seemed reasonable (at least
in part) to Einstein's contemporaries. In the science of mechanics,
not only had the principle of relativity been around since Galileo,
but for some years Poincaré (among others) had also analyzed the
relativity principle's problems and prospects in electrodynamics.[3] If
light, moreover, was nothing but an excitation of waves in a rigid,
all-pervasive ether, then in the frame of reference in which the
ether was at rest, it was plausible to assume that the speed of light
would not depend on the speed of the light source. After all, for reasonable source speeds, the speed of sound does not depend on the
velocity of the source: once a sound wave is started, it moves
through air at a fixed speed.

But how could Einstein's two postulates be reconciled? Suppose
in the ether rest frame a light was shining. To an observer *moving*
with respect to the ether, wouldn't the light appear to travel either
faster or slower than normal, depending on whether the moving
observer was approaching or retreating from the light? And if a difference in the velocity of light was observable, then wouldn't that
violate the principle of relativity, since that observation would indi-

cate whether one was truly moving with respect to the ether? Yet no such difference could be measured. Even precise optical experiments failed to detect the slightest hint of motion through the ether.

Einstein's diagnosis: "insufficient consideration" had been paid to the most fundamental concepts of physics. He claimed that if these basic concepts were properly understood, the apparent contradiction between the relativity and the light principles would vanish. Einstein proposed, therefore, to begin at the very beginning of physical reasoning, asking, What is length? What is time? And especially: What is simultaneity? Everyone knew that the physics of electromagnetism and optics depended on making measurements of time, length, and simultaneity, but as far as Einstein was concerned, physicists had not paid enough critical attention to the basic procedures by which these fundamental quantities were determined. How could rulers and clocks yield unambiguous space and time coordinates for the phenomena of the world? In Einstein's judgment, the predominant view that physicists should concern themselves first with the complex forces that held matter together had it backward. Instead, *kinematics* had to come first, that is, how clocks and rulers behaved in constant, force-free motion. Only then could the problem of *dynamics* (for example, how electrons behaved in the presence of electrical and magnetic forces) be usefully addressed.

Einstein believed that physicists would only find consistency by sorting out the measurements of space and time. To make spatial measurements, a coordinate system is needed—by Einstein's lights, a system of ordinary rigid measuring rods. For example: this point is two feet along the *x*-axis, three along the *y*, and fourteen up the *z*-axis. So far, so good. Then came the surprising part, the reanalysis of *time* that contemporaries like the mathematician and mathematical physicist Hermann Minkowski saw as the crux of Einstein's argument.[4] As Einstein put it: "We have to take into account that all our judgments in which time plays a role are always judgments of *simultaneous events*. If, for instance, I say, 'That train arrives here at 7 o'clock,' I mean something like this: 'The pointing of the small hand of my watch to 7 and the arrival of the train are simultaneous events'."[5] For simultaneity *at one point*, there is no problem: if an event located in the immediate vicinity of my watch (say, the train engine pulling up beside me) occurs just when the small hand of the watch reaches the seven, then those two events are obviously simultaneous. The difficulty, Einstein insisted, comes when we have to link events separated in space. What does it mean to say two *distant* events are simultaneous? How do I compare the reading of my watch *here* to a train's arrival at another station *there* at 7 o'clock?

For Newton the question of time held an absolute component; time was not and could not be merely a question of "common" clocks. From the instant Einstein demanded a *procedure* in order to give meaning to the term "simultaneous," he split from the doctrine of absolute time. In an apparently philosophical register, Einstein established this defining procedure through a *thought experiment* that has long seemed far from the play of laboratories and industry. How, Einstein asked, should we synchronize our distant clocks? "We could in principle content ourselves to time events by using a clock-bearing observer located at the origin of the coordinate system, who coordinates the arrival of the light signal originating from the event to be timed . . . with the hands of his clock."[6] Alas, Einstein noted, because light travels at a finite speed, this procedure is not independent of the place of the central clock. Suppose I stand next to A and far from B; you stand exactly halfway between A and B:

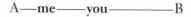

Both A and B flash light signals to me, and both arrive in front of my nose at the same moment. Can I conclude that they were sent at the same time? Of course not. It is obvious that B's signal had a much longer way to travel to me than A's signal, and yet they arrived at the same time. So B's signal must have been launched before A's.

Suppose I stubbornly insist that A and B *must* have launched their signals simultaneously; after all, I got the two signals at the same moment. Immediately I run into trouble, as you can bear witness: if you were standing exactly halfway between A and B, then you would have received B's light before A's. To avoid ambiguity, Einstein did not want to make the simultaneity of the two events "A sends light" and "B sends light" depend on where the receiver hap-

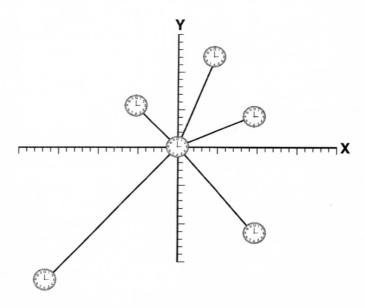

Figure 1.1 Central Clock Coordination. In his 1905 paper on special relativity, Einstein introduced — and rejected — a scheme of clock coordination in which the central clock sent a signal to all other clocks; these secondary clocks set their times when the signal arrived. For example, if the central clock sent its time signal at 3:00 P.M., each secondary clock synchronized its hands to that same 3:00 P.M. when the pulse arrived. Einstein's objection: the secondary clocks were at different distances from the center so close clocks would be set by the arriving signal before distant ones. This made the simultaneity of two clocks depend (unacceptably to Einstein) on the arbitrary circumstance of where the time-setting "central" clock happened to be.

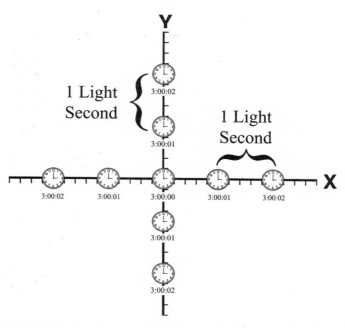

Figure 1.2 Einstein's Clock Coordination. *Einstein argued that a better and nonarbitrary solution to the simultaneity question was this: set clocks not to the time that the signal was launched, but to the time of the initial clock plus the time it took for the signal to travel the distance from the initial clock to the clock being synchronized. Specifically, he advocated sending a round-trip signal from the initial clock to the distant clock and then setting the distant clock to the initial clock's time plus half the round-trip time. In this way the location of the "central" clock made no difference—one could start the procedure at any point and unambiguously fix simultaneity.*

pens to be standing. As a procedure for defining simultaneity, "simultaneous receipt of signals by me" is a disaster, an epistemic straw man who cannot tell a consistent story.

Having knocked the straw out of this straw man, young Einstein proposed a better system: let one observer at the origin A send a light signal when his clock says 12:00 to B at a distance *d* from A;

the light signal reflects off B and returns to A. Einstein has B set her clock to noon plus half the round-trip time. A two-second round-trip? Then Einstein has B set her clock to noon plus one second when she gets the signal. Assuming that light travels just as fast in one direction as the other, Einstein's scheme amounts to having B set her clock to noon plus the distance between the two clocks divided by the speed of light. The speed of light is 300,000 kilometers per second. So if B is 600,000 kilometers from A when B receives the light signal, she sets her clock to 12:00:02, noon plus two seconds. If B were 900,000 kilometers away from A, B would set her clock to 12:00:03 when she gets the signal. Continuing in this way, A, B, and anyone else participating in this coordination exercise can all agree that their clocks are synchronized. If we now move the origin, it makes no difference: every clock is already set to take into account the time it takes for a light signal to arrive at the clock's location. Einstein liked this: no privileged "master clock," and an unambiguous definition of simultaneity.

With the clock coordination protocol in hand, Einstein had cracked his problem. By relentlessly applying the simple procedure of coordination and his two starting principles, he could show that two events that were simultaneous in one frame of reference were not simultaneous in another. Consider: the length measurement of a moving object always depends on making simultaneous position measurements of two points (if you want the length of a moving bus, it behooves you to measure the position of the front and back at the same time). Because the determination of length requires the simultaneous measurements of front and back, the relativity of simultaneity led to a relativity of lengths—my frame of reference will measure a meter stick moving by me as less than a meter long.

Astonishing in and of itself, this relativity of times and lengths led to many other consequences, some more immediate than others. Because speed is defined as distance covered in a certain time, combining the motion of objects had to be reconsidered in Einstein's theory. A person running in a train at a speed of 1/2 the speed of light (with respect to the train) while the train barreled along at 3/4 the speed of light would, in Newtonian physics, be moving relative to the ground at 1 1/4 times the speed of light. But by rigorously following the definition of time and simultaneity, Einstein showed that the actual combined speed would be less than that—indeed *always* less than the speed of light no matter what the speed of the train or the runner in the train. Nor was that all: Einstein could explain previously puzzling optical experiments and make new predictions about the motion of electrons. Finally, Einstein's starting assumptions about light speed and relativity, coupled with his clock coordination scheme, helped show that there weren't really *two different* explanations of the coil, magnet, and lamp but just one: a magnetic field in one frame was an electric field in another. The difference was one of perspective—the view from different frames of reference. And all without a whiff of ether. A short time later, Einstein was to use relativity to produce that most famous of all scientific equations, $E = mc^2$. With consequences that at first seemed restricted to the most sensitive of barely possible experiments to the utter transformation of the military-political domain forty years later, Einstein had found mass and energy to be interchangeable.

Much lies behind Einstein's relativity besides the coordination of clocks. Without exaggeration, one could say that the collective mastery of electricity and magnetism was *the* great accomplishment of nineteenth-century physical science. Theoretically, Cambridge physicist James Clerk Maxwell had produced a theory that showed light to be nothing other than electric waves and so unified electrodynamics and optics. Practically, dynamos had brought electric lighting to cities, electric trams had altered cityscapes, and telegraphs had transformed markets, news, and warfare. By the century's end physicists were making precision measurements of light—staggeringly accurate attempts to detect the elusive ether; they were refining work in electricity and magnetism to dissect the behavior of the newly accepted electron. All this led many of the leading physicists (not just Einstein and Poincaré) to consider the problem of an electrodynamics of moving bodies to be one of the most difficult, fundamental, and acute problems on the scientific agenda.[7]

By Einstein's own account, the recognition that synchronizing clocks was necessary to define simultaneity was the final conceptual step that let him conclude his long hunt, and *that*—time coordination—is the subject of this book. Indeed, Einstein judged the alteration of time in relativity theory to be that theory's most striking feature. But his assessment did not immediately carry the day, even among those who counted themselves as Einstein's backers. Some embraced relativity after experiments on the deflection of electrons seemed to lend it support. There were those who used the theory only when physicists and mathematicians had reworked it into more familiar terms that did not put so much stress on the relativity of time. Through tense meetings, exchanges of letters, articles and responses, by 1910 a growing number of Einstein's colleagues were pointing to the revision of the time concept as the salient feature. In the years that followed, it became canonical for both philosophers and physicists to hail clock synchronization as a triumph in both disciplines, a beacon of modern thought.

Younger physicists, including Werner Heisenberg, began in the 1920s to pattern the new quantum physics on what they took to be Einstein's tough stance against concepts (like absolute time) that referred to nothing observable. In particular, Heisenberg admired Einstein's insistence that simultaneity refer exclusively to clocks coordinated by a definite and observable procedure. Heisenberg and his colleagues pressed their insistence on observability hard: if you want to speak about the position of an electron, show the procedure by which that position can be observed. If you want to say something about its momentum, then display the experiment that will measure it. Most dramatically, if even in principle you could not measure both position and momentum simultaneously, then position and momentum simply did not both exist at the same time. Einstein famously bridled at that conclusion, even as his quantum colleagues pleaded with him to acknowledge that they had only extended to atoms Einstein's own acute criticism of time, and simultaneity. It was far too late for Einstein to call his relativistic genie back into the bottle, but he worried the new physics carried too far the spirit of his insistence on observable procedures — and so underestimated the formative role of theories in fixing what could be seen. As Einstein wryly observed, "A good joke should not be repeated too often."[8]

The good joke spread. Psychologist Jean Piaget made the investigation of the "intuitive" time concept in the child into an important research area. Einstein's time coordination began serving as a model — and soon *the* model — for a new era of scientific philosophy. Gathering in the Austrian capital to found a new antimetaphysical philosophy, the physicists, sociologists, and philosophers of the Vienna Circle hailed synchronized clock simultaneity as the paradigm of a proper, verifiable scientific concept. Elsewhere in Europe and in the United States, other self-consciously modern philosophers (as well as physicists) joined in hailing signal exchange simultaneity as an example of properly grounded knowledge that would stand proof against idle metaphysical speculation.[9] To Willard Van Orman Quine, one of the most influential American philosophers of the twentieth century, *all* knowledge was ultimately revisable (he even held that logic might eventually need alteration). Yet as Quine surveyed the whole of scientific understanding, he chose as most durable Einstein's definition of simultaneity through clocks and light signals, judging that it was Einstein's time concept that we "should be most inclined to preserve when called upon to make future revisions of . . . science."[10] For a philosophical century marked by vast changes in knowledge, in a climate hostile to eternal, unbending truths, there was no higher praise.

Of course not everyone admired the relativity of time. Some lampooned it, others tried to rescue physics from it. But very broadly by the 1920s, both physicists and philosophers recognized that Einstein's question, What is time? set a standard for scientific concepts that demanded something more finite, more humanly accessible than Newton's metaphysical, absolute time. Einstein himself suggested that he had drawn an effective philosophical sword against absolute time from the eighteenth-century critical work of David Hume, who had forcefully argued that the statement "A causes B" meant nothing more than the regular sequence, A then B. Key for Einstein, too, was the Viennese physicist-philosopher-psychologist Ernst Mach's work lambasting concepts disconnected from perception. Among Mach's (sometimes excessive) roundup of idle abstractions, none figured as a greater offender than Newton's "medieval" notions of absolute space and absolute time. Einstein also studied time through the microscope of other scientists' inquiries, among them those of Hendrik A. Lorentz and Poincaré. Each of these lines of philosophical reasoning—and others that we will encounter—form part of our story of time and timepieces. Yet a purely intellectual history leaves Einstein hovering in a cloud of abstractions: the philosopher-scientist brandishing thought experiments against the dusty Newtonian dogma of absolute time. Einstein confounding a contemporary scientific-technical cadre too sophisticated to ask basic questions about time and simultaneity. But is this cerebral account sufficient?

A Critical Opalescence

Certainly Einstein and Poincaré often looked back on their work as if it originated entirely outside the material world. In this respect, it is useful to reflect on a speech Einstein delivered in early October 1933 to a massive rally organized to aid refugees and displaced scholars. Scientists, politicians, and the public jammed London's

Royal Albert Hall. Hostile demonstrators threatened to stir things up; a thousand students came to serve as protective "stewards." Einstein warned of the imminence of war, of the hatred and violence looming over Europe. He urged the world to resist the drive toward slavery and oppression, and pleaded with governments to halt the impending economic collapse. Then, suddenly, the political thread of Einstein's speech snapped. There was a sudden pulling back from worldly crisis, as if the calamity of current events had stretched him beyond his limits. In a different register he began to reflect on solitude, creativity, and quiet, on moments he had spent lost in abstract thought surrounded only by the productive monotony of the countryside. "There are certain occupations, even in modern society, which entail living in isolation and do not require great physical or intellectual effort. Such occupations as the service of lighthouses and lightships come to mind."[11]

Solitude was perfect for a young scientist engaged with philosophical and mathematical problems, Einstein insisted. His own youth, we are tempted to speculate, might be thought of this way: we might read the Bern patent office where *he* had earned a living as no more than such a distant oceanic lightship. Consistent with Einstein's garden of otherworldly contemplation, we have enshrined Einstein as *the* philosopher-scientist who ignored the clutter of the patent office and the chatter of the hallway to rethink the foundations of his discipline, to topple the Newtonian absolutes of space and time. Newton to Einstein: it is easy enough to represent this transformation of physics as a confrontation of theories floating above the world of machines, inventions, and patents. Einstein himself contributed to this image, emphasizing in many places the role of pure thought in the production of relativity: "[T]he essential in the being of a man of my type lies precisely in *what* he thinks and *how* he thinks, not in what he does or suffers."[12]

The picture we so often see is of an Einstein otherworldly, oracular, communing with the spirits of physics; Einstein pronouncing on the freedom of God in the creation of the Universe; Einstein shucking off patent applications as so much busywork between him and the philosophy of nature; Einstein summoning a world of pure thought experiments featuring imaginary clocks and fantastical trains. Roland Barthes explored this imagined persona in his "Brain of Einstein," where the scientist appears as nothing but his cerebrum, an icon of thought itself, at once magician and machine without body, psychology, or social existence.[13]

Barthes would have known that among those scientists imagined to float above the material world was Henri Poincaré, the extraordinary French mathematician, philosopher, and physicist who produced, quite independently of Einstein, a detailed mathematical physics incorporating the relativity principle. In elegantly worded essays, Poincaré offered these results to the wider cultured world, at the same time probing the limits and accomplishments of both modern and classical physics. Like Einstein, Poincaré presented himself as a mind unbound. In one of the most famous accounts ever written by a scientist of his own creative work, Poincaré recounted his steps toward a theory of a new set of functions that were important for several domains of mathematics:

> For fifteen days I strove to prove there could not be any functions like those I [had in mind]. I was then very ignorant; every day I seated myself at my work table, stayed an hour or two, tried a great number of combinations and reached no results. One evening, contrary to my custom, I drank black coffee and could not sleep. Ideas rose in crowds; I felt them collide until pairs interlocked, so to speak, making a stable combination. . . . I had only to write out the results which took but a few hours.[14]

Not just in his account of his newly invented functions, but throughout his remarkable philosophical and popular essays, Poincaré dissected physics and philosophy by way of metaphorical worlds detached from the here and now, suspending imaginary scientists in idealized alternate universes: "Suppose a man were translated to a planet, the sky of which was constantly covered with a thick curtain of clouds, so that he could never see the other stars. On that planet he would live as if it were isolated in space. But he would notice that it revolves. . . ."[15] Poincaré's space traveler might exhibit the rotation by showing that the planet bulged around its equator, or by demonstrating that a free-swinging pendulum gradually rotates. As always, Poincaré here used an invented world to make a real philosophical-physical point.

It certainly is possible—even productive—to read Einstein and Poincaré as if they were abstract philosophers whose goal was to enforce philosophical distinctions by fabricating hypothetical worlds rich in imaginative metaphors. Poincaré (it might be thought) had in mind such a world when he spoke of such wildly varying temperatures that objects altered their lengths dramatically as one moved up or down. Poincaré and Einstein's attacks on Newton's absolute simultaneity could be taken to be just such metaphorical musings, ones that employed imaginary trains, fantastical clocks, and abstract telegraphs.

Let's return to Einstein's central inquiry. Invoking what may seem a quaintly metaphorical thought experiment, Einstein wanted to know what was meant by the arrival of a train in a station at 7 o'clock. I have long read this as an instance of Einstein asking a question that (as Einstein put it) was normally posed "only in early childhood," a matter that he, peculiarly, was still asking when he was "already grown up."[16] Was this the naïveté of the isolated genius? Such riddles about time and space appear, on this reading, to be so elementary as to lie below the conscious awareness of professional scientists. But was the problem of simultaneity, in fact, below the threshold of mature thought? Was no one else in 1904–1905 *in fact* asking what it meant for an observer here to say that a distant observer was watching a train arrive at 7 o'clock? Was the idea of defining distant simultaneity through the exchange of electric signals a purely philosophical construct removed from the turn-of-the-century world?

Relativity was certainly far from my mind when, not long ago, I was standing in a northern European train station, absentmindedly staring at the elegant clocks that lined the platform. They all read the same to the minute. Curious. Good clocks. Then I noticed that, as far as I could see, even the staccato motion of their second hands clicked in synchrony. These clocks are not just running well, I thought; these clocks are coordinated. Einstein must also have had coordinated clocks in view while he was grappling with his 1905 paper, trying to understand the meaning of distant simultaneity. Indeed, across the street from his Bern patent office was the old train station, sporting a spectacular display of clocks coordinated within the station, along the tracks, and on its facade.

The origins of coordinated clocks, like much in our technological past, remains obscure. Which of the many parts of a technological system does one count as its defining feature: the use of electricity? the branching of many clocks? the continuous control of the distant clocks? However one reckons, already by the 1830s and 1840s Britain's Charles Wheatstone and Alexander Bain, and soon thereafter Switzerland's Mathias Hipp and a myriad of other European and American inventors, began constructing electrical distribution systems to bind numerous far-flung clocks to a single central clock, called in their respective languages the "*horloge-mère*" [mother clock], "*Primäre Normaluhr*"[primary standard clock], and the "master clock."[17] In Germany, Leipzig was the first city to install electrically distributed time systems, followed by Frankfurt in 1859; Hipp (then director of a telegraph workshop) launched the Swiss effort in the Federal Palace in Bern, where a hundred clock faces began marching together in 1890. Clock coordination quickly embraced Geneva, Basel, Neuchâtel, and Zurich, alongside their railways.[18]

Figure 1.3 Bern Train Station *(circa 1860—65) One of the first buildings in Bern to be provided with the new coordinated clocks. Two clocks are (barely) visible just over the oval arches on the open side of the station.* SOURCE: COPY-RIGHT BÜRGERBIBLIOTHEK BERN, NEG. 12572.

Einstein, therefore, was not only surrounded by the technology of coordinated clocks, he was also in one of the great centers for the invention, production, and patenting of this burgeoning technology. Were any other major scientists whose concern was with basic physical laws of electromagnetism and the nature of philosophical time also in the midst of this vast effort to synchronize clocks? There certainly was at least one.

Some seven years before the twenty-six-year-old patent officer redefined simultaneity in his 1905 relativity paper, Henri Poincaré had advanced strikingly similar ideas. A cultured intellectual, Poincaré was widely acclaimed as one of the greatest of nineteenth-century mathematicians for his invention of a great part of topology,

Figure 1.4 Neuchâtel Master Clock. Beautifully decorated master clocks were objects of enormous value and civic pride. This one, in the center of the clockmaking region of Switzerland, received its time from an observatory and then launched its signals along telegraph lines. SOURCE: FAVARGER, L'ÉLECTRICITÉ (1924), P. 414.

his celestial mechanics, his enormous contributions to the electro-dynamics of moving bodies. Engineers lauded his writings on wire-less telegraphy. The wider public devoured his best-selling books on the philosophy of conventionalism, science and values, and his defense of "science for science."

For our purposes, one of the most remarkable essays Poincaré published appeared in January 1898 in a philosophical journal, the *Review of Metaphysics and Morals*, under the title "The Measure of Time." There Poincaré blasted the popular view, espoused by the influential French philosopher Henri Bergson, that we have an

Figure 1.5 Berlin Master Clock. *This clock, residing at the Silesischer Bahnhof in Berlin, sent its time down the many tracks emanating from the station.* SOURCE: *L'ÉLECTRICITÉ* (1924), P. 470.

intuitive understanding of time, simultaneity, and duration. Poincaré argued instead that simultaneity was irreducibly a *convention*, an agreement among people, a pact chosen not because it was inevitably in truth, but because it maximized human convenience. As such, simultaneity had to be *defined*, which one could do by reading clocks coordinated by the exchange of electromagnetic signals (either telegraph or light flashes). Like Einstein in 1905, Poincaré in 1898 contended that in making simultaneity a procedural concept, the time of transmission would have to be taken into account in any telegraphically communicated time signal.

Had Einstein seen Poincaré's paper of 1898 or a crucial subsequent one of 1900 before he wrote his 1905 paper? Possibly. While there is no definitive evidence one way or the other, it will, nonetheless, prove worthwhile to explore the question both narrowly and more widely. For as we will see, Einstein need not have read just those lines of Poincaré. Clock coordination appeared in the pages of philosophy journals, and even occasionally in physics publications. In fact, electromagnetic clock coordination was so fascinating to the late-nineteenth-century public that the subject came in for close discussion in one of Einstein's favorite childhood books on science.[19] In 1904–05, clock-coordinating cables were thick on the ground and under the seas. Synchronized timepieces were everywhere.

Just as commentators have grown used to interpreting Einstein's talk of trains, signals, and simultaneity as an extended metaphor, a literary-philosophical thought experiment, there is a similarly routine metaphorical reading of Poincaré's observations. Here too, supposedly, stands philosophical speculation, an anticipatory note to Einstein's special theory of relativity, a brilliant move by an author lacking the intellectual courage to pursue it to its logical, revolutionary end. So familiar is this story that it has become a commonplace to treat Poincaré's insight into coordinated time as if it were entirely isolated, a philosophical *aperçu* disconnected from his place in the world. But neither Poincaré nor Einstein was speaking in a vacuum about time.

What, Poincaré asks, are the rules by which scientists judge simultaneity? What *is* simultaneity? His final, most forceful example turned on the determination of longitude. He began by noting that when sailors or geographers determine longitude, they must solve precisely the central problem of simultaneity that governed Poincaré's essay: they must, without being in Paris, calculate Parisian time.

Finding latitude is simple. If the north star is straight overhead, you are on the North Pole; if it is halfway to the horizon, you are at the latitude of Bordeaux; if it is on the horizon, you are at the latitude of Ecuador, on the equator. It does not matter at all what time you make latitude measurements—in any particular location the angle of the pole star is always the same. Finding the longitude difference between two points is famously more difficult: it requires two distant observers to make astronomical measurements *at the same time*. If the earth did not rotate, there would be no problem: you and I would both look up and check which stars were directly under the North Star (for example). By checking a map of the stars we could easily find our relative longitudes. But of course the earth does turn, so to fix longitude differences accurately we must be sure that we are measuring the position of the overhead stars (or sun or planets) at the same time. For example, suppose a map-making team in North America knew the time in Paris and saw that at the team's location the sun rose exactly six hours later than it had in the City of Light. Since the earth takes 24 hours to rotate, the team would know that it was somewhere along a longitude line 6/24 (one-fourth or equivalently 90 degrees) of the way around the world to the west of Paris. But how could the explorers know what time it was back in Paris?

As Poincaré says in his "Measure of Time," the roving cartographer could know Paris time simply by carrying a precision timekeeping device (chronometer) on the expedition, having set it to Paris time. But transporting chronometers led to problems both in principle and in practice. The explorer and his Parisian colleagues could observe an instantaneous celestial phenomenon (such as the emergence of a moon of Jupiter from behind the planet) from their two different locations and declare that their observations were simultaneous. But this seemingly simple procedure isn't. There were practical problems in using Jovian eclipses. Even as a matter of principle, as Poincaré noted, the time would need to be corrected because light from Jupiter travels over different paths to reach the two observation points. Or—and this is the method Poincaré pursues—the explorer could use an electric telegraph to exchange time-signals with Paris:

It is clear first that the reception of the [telegraph] signal at Berlin, for instance, is after the sending of the same signal from Paris. This is the rule of cause and effect. . . . But how much after? In general, the duration of the transmission is neglected and the two events are regarded as simultaneous. But, to be rigorous, a little correction would still have to be made by a complicated calculation; in practice it is not made, because it would be well within the errors of observation; its theoretic necessity is none the less from our point of view, which is that of a rigorous definition.[20]

Direct intuitions about time, Poincaré concluded, are incompetent to settle questions of simultaneity. To believe so is to fall into illusion. Intuitions must be supplemented by rules of measurement: "No general rule, no rigorous rule; a multitude of little rules applicable to each particular case. These rules are not imposed upon us by themselves, and we might amuse ourselves in inventing others; but they could not be cast aside without greatly complicating the laws of physics, mechanics, and astronomy. We choose these rules, therefore, not because they are true, but because they are the most convenient."[21] All these concepts—simultaneity, time order, equal durations—were defined to make the expression of natural laws as simple as humanly possible. "In other words, all these rules, all these definitions are nothing but the fruit of an unconscious opportunism."[22] Time, according to Poincaré, is a *convention*—not absolute truth.

What time do the map makers make it out to be in Berlin when it is noon in Paris? What time is it down the line when the train pulls into Bern? In posing such questions, Poincaré and Einstein seem, at first glance, to be asking questions of stunning simplicity. As was their answer: two distant events are simultaneous if coordinated clocks at the two locations read the same—noon in Paris, noon in Berlin. Such judgments were inevitably *conventions* of procedure and rule: to ask about simultaneity was to ask how to coordinate clocks. Their proposal: Send an electromagnetic signal from one clock to the other, taking into account the time the signal takes to arrive (at approximately the speed of light). A simple idea of breathtaking consequences for concepts of space and time, for the new relativity theory, for modern physics, for the philosophy of conventionalism, for a world-covering network of electronic navigation, for our very model of secure scientific knowledge.

This is my quarry: how, at the turn of the century, was simultaneity actually produced? How did Poincaré and Einstein both come to think that simultaneity had to be defined in terms of a conventional procedure for coordinating clocks by electromagnetic signals? Addressing these questions demands far too wide a scope to be captured by a biographical approach, though there are, to be sure, too many biographies of Einstein and not enough of Poincaré. Nor is this book a history of philosophical ideas of time, a task that could easily take us back before Aristotle. It is not a comprehensive account of the intricate development of timepieces, even electric ones. And it is not a complete history of the many broadly shared concepts of nineteenth-century electrodynamics that Poincaré and Einstein appropriated as each struggled to reformulate the electrodynamics of moving bodies.

Instead, this is a slice through layers of physics, technology, and philosophy that cuts high and low, an exploration of synchronized clocks crisscrossing back and forth between the wiring of the oceans to marching Prussian armies. It reaches into the heartland of physics, through the philosophy of conventionalism, and back through relativistic physics. Take hold of a wire in the late-nineteenth-century telegraph system and begin to pull: it takes take you down and across the North Atlantic, up onto pebbled beaches of Newfoundland; it tracks from Europe into the Pacific and up into Haiphong Harbor; it slides along the ocean floor the length of West Africa. Follow the land-based wires and the iron and copper cables; they lead up into the Andes, through the backcountry of Senegal, and clear across North America from Massachusetts to San Francisco. Cables run along train lines, under oceans, and between the beachfront shacks of colonial explorers and the chiseled stone of great observatories.

But wires for time did not arrive on their own. They came with national ambitions, war, industry, science, and conquest. They were a visible sign of the coordination among nations in conventions about lengths, times, and electrical measures. Coordinating clocks in the nineteenth and twentieth centuries was never just about a little procedure of signal exchange. Poincaré was an administrator of this global network of electrical time, Einstein an expert at the central Swiss clearinghouse for the new electrotechnologies. Both were also riveted by the electrodynamics of moving bodies and fascinated by philosophical reflections on space and time. Understanding this world-embracing synchronization will take us some way toward understanding what is modern about modern physics and about how Einstein and Poincaré stood at crossing points of their respective modernities.

Surely, we learn from the astonishing contrast between Newton's distant seventeenth- and Einstein and Poincaré's turn-of-the-twentieth century concepts of time. Their two conceptions stand as monuments to a clash between the early modern and the modern: on the one hand, space and time as modifications of the sensorium of God; on the other, space and time as given by rulers and clocks. But the distance between 1700 and 1900 should not eclipse the near at hand. It is the near at hand that interests me—the daily world of 1900 in which it became usual, and not just for Poincaré and Einstein, to see time, conventions, engineering, and physics as of a piece. For those decades it made perfect sense to mingle machines and metaphysics. A century later that propinquity of things and thoughts seems to have vanished.

Perhaps one reason for the difficulty we have in imagining science and technology so caught up with one another is that it has become habitual to divide history into separate scales: intellectual history for ideas that are or aim to be universal; social history for more localized classes, groups, and institutions; biography or microhistory for individuals and their immediate surround. In telling of the relation between the pure and applied, there are narratives that track abstract ideas down through laboratories to the machine-shop floor and into everyday life. There are also stories that run the other way, in which the daily workings of technology are slowly refined as they shed their materiality on the way up the ladder of abstraction until they reach theory—from the shop floor to the laboratory to the blackboard, and eventually to the arcane reaches of philosophy. Indeed, science often does function this way: from the purity of an etherial vapor, ideas may seem to condense into everyday matter; conversely, ideas seem to sublime from the solid, quotidean world into air.

But here neither picture will do. Philosophical and physical reflections did not *cause* the deployment of coordinated train and telegraph time. The technologies were not derivative versions of an abstract set of ideas. Nor did the vast networks of electro-coordinated clocks of the late nineteenth century cause or force the philosophers and physicists to adopt the new convention of simultaneity. No, the present narrative of coordinated time fits neither of these metaphors of progressive evaporation or condensation. Another image is needed.

Imagine an ocean covered by a confined atmosphere of water vapor. When this world is hot enough, the water evaporates; when the vapor cools, it condenses and rains down into the ocean. But if the pressure and heat are such that, as the water expands, the vapor is compressed, eventually the liquid and gas approach the same density. As that critical point nears, something quite extraordinary occurs. Water and vapor no longer remain stable; instead, all through this world, pockets of liquid and vapor begin to flash back and forth between the two phases, from vapor to liquid, from liquid to vapor—from tiny clusters of molecules to volumes nearly the size of the planet. At this critical point, light of different wavelengths begins reflecting off drops of different sizes—purple off smaller drops, red off larger ones. Soon, light is bouncing off at every possible wavelength. Every color of the visible spectrum is reflected as if from mother-of-pearl. Such wildly fluctuating phase changes reflect light with what is known as critical opalescence.

This is the metaphor we need for coordinated time. Once in a great while a scientific-technological shift occurs that cannot be understood in the cleanly separated domains of technology, science, or philosophy. The coordination of time in the half-century following 1860 simply does not sublime in a slow, even-paced process from the technological field upward into the more rarified realms of science and philosophy. Nor did ideas of time synchronization originate in a pure realm of thought and then condense into the objects and actions of machines and factories. In its fluctuations back and forth between the abstract and the concrete, in its variegated scales, time coordination emerges in the volatile phase changes of critical opalescence.

To dig into the records of almost any town in Europe or North America—indeed, far beyond both—reveals the struggle to coordinate time during these years of the late nineteenth century. There lie the yellowing data of railroad superintendents, navigators, and jewelers, but also of scientists, astronomers, engineers, and entrepreneurs. Time coordination was an affair for individual school buildings, wiring their classroom clocks to the principal's office, but also an issue for cities, train lines, and nations as they soldered alignment into their public clocks and often fought tooth-and-nail over how it should be done. Step back to the archives of central governments and the cast of characters grows wider and wilder: anarchists, democrats, internationalists, generals.

Amidst this cacophony of voices, this book aims to show how the synchronizing of clocks became a matter of coordinating not just procedures but also the languages of science and technology. The story of time coordination around 1900 is not one of a forward march of ever more precise clocks; it is a story in which physics, engineering, philosophy, colonialism, and commerce collided. At every moment, synchronizing clocks was both practical and ideal: gutta-percha insulator over ironclad copper wire *and* cosmic time. So variously construed was time regulation that it could serve in Germany as a stand-in for national unity, while in France at the same moment it embodied the Third Republic's rationalist institutionalization of the Revolution.

My aim is to pursue coordinated time through this critical opalescence, and in particular to set Henri Poincaré and Albert Einstein's revamped simultaneity in the thick of it. Entering the sites of time production and the lanes of its distribution will bring us repeatedly to two crucial locations in the binding of clocks that joined Einstein's and Poincaré's transcendent metaphors of clocks and maps to altogether literal places: the Paris Bureau of Longitude and the Bern Patent Office. Standing at those two exchanges, Poincaré and Einstein were witnesses, spokesmen, competitors, and coordinators of the cross-flows of coordinated time.

Order of Argument

Because the fate of coordinated time cannot be tracked from a nuclear group of railroad managers, inventors, or scientists in a simple widening circle, our story will switch scales back and forth between local and global narratives. I want to introduce Poincaré in chapter 2 ("Coal, Chaos, and Convention") in a way that may be somewhat unfamiliar. Who would guess from *Science and Hypothesis*, his best-selling book of 1902, that he had trained as a mining engineer and served as an inspector in the dangerous, hard-pressed coal mines of eastern France? Or that for decades he had helped run the Bureau des Longitudes in Paris, serving as its president in 1899 (and later in both 1909 and 1910)? Or that he co-edited and often published in a major journal on electrotechnology that ran abstract articles on fundamental issues of electrodynamics next to pieces about undersea cables and the electrification of cities?

Understanding the transformation of time—its radical secularization—requires a relocation of Poincaré, whose conventionalization of simultaneity is flattened into two dimensions if he is seen purely as a mathematician-philosopher or mathematical physicist (though he was certainly both). More is needed than a simple addition of a side-interest in engineering. Poincaré figures here not as a free-floating monad who seized this or that "resource" from philosophy, mathematics, or physics to solve particular problems. Instead, throughout this book I want to situate Poincaré in the midst of powerful series of moves that, at a few critical intersections, formed roughly consistent ways of acting within physics (or philosophy or engineering). It is not that a preformed Poincaré plucked mere details of procedure from his alma mater, Ecole Polytechnique, but instead that he was also very much its product. As he put it, he and his colleagues proudly bore Polytechnique's "factory stamp." Chapter 2 is about that stamp, under which it was as reasonable for Poincaré to assess mining accidents as to augur the stability and fate of the Solar System or to generate abstract mathematics. It takes us close-in to Poincaré to capture this wider link between the material and the abstract. And that bond is crucial if we are to understand in subsequent chapters the multiple ways in which Poincaré insisted on examining the concept of simultaneity under the different but intersecting lights of physics, philosophy, and technology.

But the "factory" of Poincaré's Polytechnique education—along with the mining and mathematical years that followed—is still not yet a wide enough terrain on which to locate the secularization of time coordination. That larger territory extended even beyond France, to the colliding networks of wires and rails that the Great Powers were erecting at breakneck speed. Matching these systems at their often-contested boundaries could only be accomplished by codes and conventions that, in the 1870s and 1880s, aimed, sometimes painfully, to mesh the collision of incompatible length and time standards. So in chapter 3 we will pull back from the close-in frame of chapter 2.

Chapter 3, "The Electric Worldmap" is after this vastly larger timescape of earth-covering networks—clashing empires of time. In those late-nineteenth-century decades, nowhere was the demand for world-covering conventions more apparent than in the drawing of global maps. Confronting a staggering increase in the volume of trade during these years, navigators were increasingly frustrated by maps with differing and often unreliable longitudinal grids. So too were colonial authorities, as they stepped up the pace of the conquest of new lands, the exploitation of resources, and the building of railroad lines. All demanded precise and consistent geodesy. These various demands came to a head in an 1884 meeting at the American State Department, when twenty-two countries fought their way toward a single prime meridian—the zero point of longitude—to be placed in Greenwich, England. Frustrated, indeed infuriated by the arrogation of that global zero-arc to the hub of the British Empire, the French delegation lobbied for a decimalized time, determined to put a French stamp of rational enlightenment on the new world order of clocks and maps.

Chapter 4, "Poincaré's Maps," picks up at an intermediate range, at the height of this French campaign of the 1890s to rationalize time—in which Poincaré played a determinative role. Charged with the evaluation of the longstanding French revolutionary proposal to decimalize time, and, with it the divisions of the circle, Poincaré and his interministerial committee encountered firsthand the competing proposals for how to conventionalize the measure of time. Indeed, it was precisely in this period that Poincaré began serving as a senior member of the French Bureau of Longitude, the agency charged with coordinating clocks around the world to produce the most accurate of maps. Here, in Europe, Africa, Asia, and the Americas, in the world of precision-synchronized times and geo-

desic maps, we can finally confront Poincaré's philosophical proposal of 1898 to treat simultaneity as a convention. If simultaneity
could only be specified by agreement on how to synchronize clocks,
then a felicitous precedent would be to coordinate them exactly as
did the telegrapher-longitude finders. This move—at once a statement of up-to-date cartography and of the metaphysics of time—is
of extraordinary importance. Newton's absolute, theological time
had no place; instead there stood a *procedure*. Engineering common time stood where God's absolute time had been.

Nothing in Poincaré's 1898 conventionalization of time directly
addressed either electrodynamics or the principle of relativity. That
connection came only in December 1900, as Poincaré reexamined
earlier work of the Dutch physicist, Hendrik Antoon Lorentz. Back
in 1895, Lorentz had advanced a theory of the electron that incorporated the following extremely clever idea. In the rest frame of the
ether, where the equations governing electric and magnetic fields
(Maxwell's equations) were supposed to hold good, Lorentz spoke of
"true time," t_{true}. Suppose some object, such as an iron block, were
moving in this ether rest frame (traveling through the ether) and that
Maxwell's equations gave a detailed description of the electric and
magnetic fields in and around the iron. How should the physics be
described from a frame traveling *with* the iron block? It seemed as if
the physics would suddenly become vastly more complicated as one
tried to take account of the fact that the moving frame was racing
through the ether. But Lorentz found that he could simplify the equations, making them as simple as in the ether rest frame, if he redefined the fields and the time variable. Because he had redefined the
time of an event so that it depended on *where* the event took place,
Lorentz called t_{local} "local time" (*Ortszeit*), the same word used in
everyday life to describe the (longitude-dependent) time of Leiden,
Amsterdam, or Djakarta. The crucial point was this: Lorentz's local
time was purely a mathematical fiction used to simplify an equation.

Poincaré first published a time paper in January 1898 in a philosophy journal. His goal was to show that clock coordination by means of electric telegraph exchange formed the basis for a conventional definition of simultaneity. It was technological and philosophical and had strictly nothing to do with the physics of moving bodies. By contrast, in his second move (1900), Poincaré dramatically extended Lorentz's t_{local} to the physics of altogether real (not mathematical) moving frames of reference. True, Poincaré did everything possible not to call attention to the difference between his "apparent" local time and Lorentz's mathematical one. Nonetheless, the concept had moved: in Poincaré's hands, local time lost its fictional status and became the time that moving-frame observers would show on their clocks as they corrected for the fact that signal exchange would beat against or run with an ether wind.

With Poincaré's 1900 interpretation of local time, suddenly all three series—physics, philosophy, and geodesy—intersected in the coordination of electrosynchronized clocks. Again responding to Lorentz, Poincaré made his third move with synchronized clocks in 1905–06. In 1904, Lorentz had modified his local time, t_{local}, to make the equations of electrodynamics in the fictional moving frames even more similar to the "true" ether rest frame. Poincaré seized Lorentz's result, adjusting (inter alia) the definition of local time to make the mathematical correspondence exact between fictional moving frames and the true resting one. But the crucial point for Poincaré was not that he had slightly modified Lorentz's theory. Rather, it was this: Poincaré had demonstrated that clocks coordinated as they moved through the ether would give exactly Lorentz's new local time and would do so for real observers moving in that frame. The relativity principle held good, even while Poincaré continued to oppose "apparent time" to "true time." By 1906, Poincaré had placed the light-coordination of clocks front and center for three projects fundamental to modern knowledge: technology, philosophy, and physics.

The French polymath had begun with geodesic time, shifted registers to his antimetaphysical, conventional time, and then cleared his way to the physics of local time and relativity. Throughout his physics—but also throughout his philosophy, technology, and politics—Poincaré saw his world as improvable through rational, intuitive intervention; he reveled in pushing a problem into "crisis" and then resolving it. His was the optimism of a progressive engineer willing to reassemble major struts and cables of the structures he tackled, but insisting that the built world of "our fathers" be respected, incorporated, improved.

In chapter 5, we turn to "Einstein's Clocks." Not to the prophetic, world-famous, and mathematically inclined Einstein of 1933 or 1953, but to Einstein the tinkerer, souping up homebuilt instruments in his Kramgasse apartment, the Einstein riveted by the design of machines and the analysis of patents. This was neither the suddenly celebrated Einstein of post-World War I Berlin nor the abstracted, hermitlike elder Einstein of Princeton, but the thoroughly engaged 1905 youth of Bern. Though its technological infrastructure came late, when Switzerland inaugurated its rail, telegraph, and clock network, synchronized time there was a very public affair—and Bern was its center. From Bern electric time radiated outward, to the clock industry of the Jura region, to the public display of urban clocks, to the railroad, and, of course, to the patenting of synchronized clocks. Einstein was in the thick of it.

Yet Einstein's path to coordinated time was very different from Poincaré's. Einstein's vision was less ameliorist. Framing himself as a heretic and an outsider, Einstein scrutinized the physics of the fathers not to venerate and improve, but to displace. Einstein saw the coordination of time, and indeed his physics and his philosophy more generally, as part and parcel of the same critical reevaluation of the founding assumptions of the disciplines. Einstein did not move methodically from one aspect of time coordination to another. Most elements of his approach to relativity theory were already in place before he even touched the problem of time; for example, by 1901 he had already dismissed the ether that Poincaré had struggled so hard to maintain. Einstein's engagement with critical works at the boundary of physics and philosophy were by then long-standing, and at the patent office, he and his fellow inspectors had been dissecting the machinery of time for three years. So in May 1905, when Einstein began defining simultaneity by way of electrocoordinated clocks, it was not, as it was for Poincaré, a matter of distinguishing "apparent" from "true" time alongside a quasi-fictional ether. For Einstein clock coordination was the turn of the key that would at last set in motion the theory machine that he had struggled for a decade to assemble. There was no ether; there were actual fields and particles, and there were but real times given by clocks.

In the final chapter, "The Place of Time," it is possible to see how the work of Einstein and of Poincaré could be both extraordinarily close yet far apart. Precisely in the uneasy relation between their competing uses of coordinated time, they stand not as progressive against reactionary approaches to nature, but as two strikingly different visions of what made modern physics modern—the consummate Polytechnician's hopeful reforms against an outsider's rebellion against foundations. Yet despite their differences, both were grappling with the same extraordinary insight into electrocoordinated time, and in so doing both men stood at the crossing of two great movements. On one side lay the vast modern technological infrastructure of trains, shipping, and telegraphs that joined under the signs of clocks and maps. On the other, a new sense of the mission of knowledge was emerging, one that would define time by pragmatism and conventionality, not by eternal truths and theological sanction. Technological time, metaphysical time, and philosophical time crossed in Einstein's and Poincaré's electrically synchronized clocks. Time coordination stood, unequaled, at that intersection: the modern junction of knowledge and power.

Chapter 2

COAL, CHAOS, AND CONVENTION

ITH THE DRAMATIC success of Poincaré's 1902 collection of philosophical-scientific articles, *Science and Hypothesis*, his position in French intellectual life was unmatched. Standing at the pinnacle of mathematics, physics, philosophy, he was scientific rapporteur of a major scientific expedition and had served as president of the Bureau of Longitude. He embodied the successful Polytechnician, garnering an early reputation in mathematics and then proceeded directly to the nitty-gritty of mining engineering. Here is our central theme—intertwined abstraction and concreteness—far more closely bound for Poincaré and his contemporaries than scholars and engineers of later times ever expected. No wonder that the former students of Ecole Polytechnique asked him to use their annual address to reflect "On the Part Played by Polytechnicians in the Scientific Work of the XIXth Century." As he prepared that 25 January 1903 lecture, Poincaré sifted through the presentations of recent honorees. "I see among my predecessors," he told his classmates and colleagues, the "dean of our generals, numerous ministers, scientific engineers, the director of a large company, those who conquered our overseas empire and those who organized it, and those who brought from the earth, three years ago, the ephemeral splendors of the [Universal Exposition on the] Champ de Mars." How, Poincaré asked, could a single education have produced such a combination of scientists, soldiers, and engineering entrepreneurs?[1]

At a time when the division of labor had penetrated science, Poincaré wondered aloud what the Ecole had done to hold together those disparate specializations. True, he allowed, the competitive exams selected gifted students with diverse aptitudes; true too that the traditional "nobility" of the school served as an inspiration ("Those who fight for their country, like those who combat for truth. . . .") But something more must be at work to produce a worldview that pervaded such scattered pursuits. Perhaps it was a rubbing of shoulders—certainly chemists, physicists, and mineralogists all benefitted from the high mathematical culture of their formative schooling. Even among the most abstract of Polytechnique mathematicians, moreover, "we see the constant concern with applications"—not just among the obvious figures who had made applications their metier, but also among the most scholarly of the school's products, including one of the most powerful mathematicians in the history of Polytechnique, Augustin Louis Cauchy. Cauchy, despite his well-known opposition to a socially engaged mathematics, drew strongly on mechanics. Poincaré then invoked his own teacher, Alfred Cornu, who liked to repeat that mechanics was the cement that held together the various parts of the Polytechnician's soul. "There it is," he concluded, "that factory stamp that I was looking for; our physicists, our mathematicians are all a bit mechanicians."[2]

If the Polytechnicians—Poincaré foremost among them— emerged from their training with the "factory stamp" of mechanics imprinted on their outlook, this already differentiated them from scientists trained in the universities. But for Poincaré that was not all. University professors fretted about the unity of science; Polytechnicians had another concern, that of joining thought with action. It was *action*, Poincaré insisted, that inoculated Polytechnique's graduates against the melancholy that assailed many university researchers over the purported failure of science. That bond between the abstract and the concrete was the *most* essential feature of the "factory." Science changed; so too did the world. Poincaré was fully prepared to allow the education to evolve: "The Ecole must transform itself bit by bit like all things human, but it must not touch that which makes its soul, the alliance of theory with practice must not be broken; it must not be mutilated, for without it there would only remain a hollow name."[3]

When Poincaré had come to Polytechnique back in 1873, the problem of balancing pure knowledge against useful applications was gripping France as never before. Just two years earlier, Germany had pounded France into a humiliating defeat, leaving Germany with new lands, united, euphoric, building scientific institutes and monuments to hail the victory.[4] France, now minus Alsace-Lorraine, struggled desperately to understand the roots of its catastrophic defeat. The technical infrastructure of the State came into question: book after book lamented the sorry state of the country's railroads and, more generally, its inadequate preparation for a new age of fighting. But more than any particular piece of technology, critics dissected the institutions of technical learning, which appeared to need reconstruction, and fast. None of these institutions had greater weight than Ecole Polytechnique, the *Grande Ecole* Poincaré had just entered.

Polytechnique, founded in 1794, had no parallel in the United States, Britain, or Germany. Built to train an elite corps of engineers able to mold the military into a force of modernity by using an enlightened science, Polytechnique was to be a scientific institution where mathematics, physics, and chemistry would prosper with mathematics above all. Its students chosen by a competitive examination (the famous *concours*), Polytechnique produced a fraternity rigorously schooled in a mathematics-based engineering. After their graduation, generations of these students joined the highest levels of the State's administrative structure, overseeing at first the new nation and, later, in the nineteenth century, an empire. Part

Oxbridge and part Sandhurst? Half MIT, half West Point? Such comparisons fail. The Polytechnician was far more scientific than his British equivalent schooled in the Classics, more mathematical than a rising American engineer, and less inclined than an elite German physics student toward precision laboratory work. Polytechnique was, and remains, a unique institution, possessing by the 1870s a mythological status in France.

For the first years after its founding, the revolutionary engineer and mathematician Gaspard Monge used geometry to successfully bind the conflicting demands of practical engineering and advanced mathematics. Projective geometry, he believed, trained the mind, revealed scientific truth, and sculpted stone, wood, and fortifications. Over the course of the early decades of the nineteenth century, however, Monge's program began to falter. Mathematics, significantly propelled by the conservative Cauchy, self-consciously began to shed its ambition of joining the world with the mind. Science ascended, applications retreated—a trend encouraged by the construction of new, more specialized engineering schools.[5]

When Paris fell to the Prussians, the French found no shortage of institutions to blame. Pasteur, joined by a chorus around France, lent his prestigious voice to those who saw the German triumph as a failure of French science. At Polytechnique the charges echoed loudly—it was a major center of scientific engineering and the students were already in uniform. Alfred Cornu, a former student of the school who had returned as a rising star among the physicists, articulated the school's position after the disaster, one delicately balancing science and worldly engagement. He became, so to speak, an ideal type of the new Third Republic Polytechnician, gliding easily from the pure to the applied. As Poincaré later remarked with admiration, Cornu's work traversed a wide field of optics, bringing new instruments and techniques not only to physicists but also to astronomers, meteorologists, and even clockmakers. Having designed and built at Nice an extraordinarily precise astronomical clock with a massive pendulum, Cornu traveled there every year to perfect his clock. In addition, Cornu developed a detailed mathematical theory of the coordination of clocks by the exchange of electric signals.[6] Neither his joining of the pure to the applied nor his specific interest in electric time coordination was lost on Poincaré.

Cornu's course was filled with stylized demonstrations of particular laboratory phenomena. Although he did not train his students to do the handwork themselves, experiments were for Cornu a vital part of science — its alpha if not its omega. Nor was Cornu alone: he and his colleagues made sure that a young scientist emerging from Polytechnique in the 1870s did so with a deep respect for experiments (though no great familiarity with the manipulation of devices, complex measurements, or data analysis). Instead, students learned to see the great mathematical structures that embraced data as the endpoint of scientific work. They learned to put little faith in the literal truth of particular assumptions about atomism or electric fields. It is telling that when Polytechique filled a position in chemistry, the faculty aimed above all to maintain the balance between those who believed in atoms and those who did not.[7]

This was the world that Poincaré entered in November 1873. Competitive, alert, engaged, Poincaré watched his fellow students like a hawk, writing home about his rivals' grades, occasionally commenting on student hazings or musing over the political machinations of the Jesuits. Mechanics seized his attention, and in this all-important subject his grades rose rapidly toward nineteen or twenty (out of a possible twenty). With pride, he wrote home that he had found a simpler demonstration than the one his professor had presented in class. He wrote, too, of his progress in both technical and freehand drawing. When, to the surprise of the workers, he and a group of Polytechnique students visited a local crystal factory, Poincaré, told of his fascination with the workers' dexterity and the technology of their Siemens ovens.[8] Even as a student, Poincaré recognized the "factory stamp" of Polytechnique. But in the account reported to his mother, the teaching held none of its later glow:

> It is as if we are in an immense machine whose movement we must follow on pain of being passed by; we have to do what 20 generations of X [Polytechnique] did before us and what $2n + 1$ generations of conscripts will do after us. Here you use only two faculties of your intelligence: memory and elocution; understanding a course is something anyone can do with some work and that is why everyone can get me to cram if they want. . . . I am therefore condemned to this choice: give up personal work or stay in my place; as this alternative lasts only two years, the choice I have made is not in doubt; because the advantage I will take from my position is incommensurable; but it has to be guarded and [switching to English:] this is the question.[9]

On Cornu's death, Poincaré spoke, quite personally, of how much Cornu would be mourned, as a friend, an advisor, and as a master teacher—and Poincaré's career followed in many respects a similar path. Both had been star students at Polytechnique, both had gone on to the Ecole des Mines, both accepted the call back to Polytechnique to teach, both served the State in engineering-administrative capacities. The lessons stamped on them were deep. Poincaré, Cornu, and their contemporary Polytechnicians maintained a lifelong commitment to the link between abstract and concrete knowledge. It was a testimony to their training that both served on the board of the Bureau of Longitude and together along with their fellow Polytechnicians, drove a myriad of technologically progressive projects, from the electrotechnical journal they helped run to the scientific commissions on which they served.

Of course, the relation between pure science and engaged technology was not frozen once and for all at Polytechnique—or anywhere else. It was much more like a vast and slowly undulating sea, sometimes with olympian science looming high above a trough of technology; at other times with a triumphalist military and industrial technology cresting over the claims of pure knowledge. For a brief moment after the great defeat of 1871, Cornu and his allies flattened this sea into a smoother surface of roughly equivalent claims, at least around Polytechnique. Inside this "immense machine" where technology and mathematical physics circulated with equal force, Poincaré became Poincaré.

Coal

Before turning to clocks it is worth exploring two moments early in Poincaré's career, for they tell us a great deal about who Poincaré was, how he thought, and where he stood in the turbulent flux of technologies and sciences. As a shorthand, we may think of these two episodes as moments of coal and chaos, for during the first decade of his work—from the late 1870s to 1890 or so—Poincaré was grappling not only with a new, wildly unstable mechanics of the Solar System but also with the grubby, dangerous world of mining in late-nineteenth-century France.

Graduating in 1875, Poincaré followed tradition by moving with the other two valedictorians of his class to the Ecole des Mines. There, the three—Poincaré, Bonnefoy, and Petitdidier—began their studies in October.[10] Poincaré's mentor, mathematician Ossian Bonnet, tried to get Poincaré's course load reduced because of his mathematical work, but the School of Mines would hear nothing of it. So Poincaré finished his mathematical thesis while learning the ins and outs of ventilator shafts. Geological fieldwork took him, for example, to Austria and Hungary in 1877.

Poincaré's cultural circle was also expanding. His adored younger sister, Aline, introduced him to the philosopher Emile Boutroux, whom she would eventually marry. Boutroux and Poincaré immediately began discussing philosophy. Having studied in Heidelberg

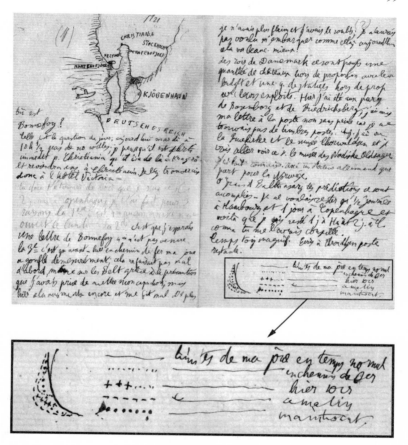

Figure 2.1 *Poincaré's Curve of Happiness* (**with detail**). *This letter, written by Poincaré to his mother during the summer of 1879, includes both a hand-drawn map of his geological travels and a geometrical curve showing the "limits of his joy" in "normal times," "yesterday," "in trains," and "right now."*
SOURCE: ARCHIVES HENRI POINCARÉ, M021.

until the Franco-Prussian War, Boutroux had made his own the German philosophical commitment to join the humanities and the sciences. Productive, enthusiastic, religiously inclined, Boutroux argued (following the Kantianism he imbibed in Heidelberg) that much in the scientific domain was more of the mind than "out there." Through Boutroux, Poincaré came to meet other philosophers, including the philosophically inclined mathematician Jules Tannery. The group did not share Boutroux's religiosity (Tannery was an ardent, secular Republican), but all sought a middle way between science as simple observation and science as mental creation. Poincaré seemed to find the view congenial; for years he, too, argued that science needed just such a mix of induction and deduction. As Poincaré put it in a letter to Aline sometime around 1877, observation and induction ought be treated with "reserve." He added, "you will say induction can only give us knowledge of the same nature as the observations themselves; observation cannot teach us anything about Substance . . . it can only show us the phenomenon by itself; and not even the phenomenon in itself; but only the sensations that it produces in us." Experience, Poincaré told his sister, could never be sufficient to ground the full generality of knowledge. "Eh what do you want me to do about it; let us always take what belongs to us, and as for the rest, we must resign ourselves to admit that it will remain to us forever but a dead letter."[11]

Another of Poincaré's fellow Polytechnicians, the philosopher-scientist Auguste Calinon, held a similarly cautious view about knowledge that "did not belong to us." In 1885, Calinon published a treatise on the foundations of mechanics and geometry. He and Poincaré seemed to have been on good terms; when they saw each other in early August 1886, Calinon followed up with a copy of his recent *Critical Study of Mechanics*. It began, straightaway, with a cautionary note about absolutes in space and time:

> Many authors, in mathematics as in philosophy, accept the notion of absolute movement as a first principle (idée première). This opinion has been much contested; . . . it is worth remarking that from the point of view of rational mechanics, this question has no importance; the movement of a point, considered in isolation, is a

purely metaphysical conception, because, even allowing that one could imagine such a movement, it is impossible to certify it and to determine its geometric conditions, for example, the form of its trajectory.[12]

Because of the inaccessibility of the absolute, Calinon would only speak of relative movement. Similarly, he suggested that simultaneity too had to be accessible: two moving stellar objects at particular positions would be called simultaneous only if one saw them "at the same time." For Calinon, the human registration of the events was so important that he wanted to take into account the time it took for the sensations to be registered by the brain: "The very idea of time is therefore inherent to the mode by which our brain functions and has no sense except for minds made like ours."[13] A Kantianism, no doubt, but one that was distinctly more psychological (or psychophysiological) than that which dominated the German-speaking scene. Apparently Poincaré wrote back almost immediately addressing the work point-by-point. Though that letter is lost, Calinon's reply survives, from which it is clear that Poincaré's very first comment addressed the notion of "at the same time." It appears that Poincaré agreed that "it is by our sensations alone that we judge simultaneity or successiveness."[14]

Poincaré's philosophical speculations about the limits of scientific knowledge, the restricted power of observation, and the active role that the mind must play in making science, were themes that remained with him for the rest of his life. But none of these metaphysical musings interrupted either his mathematical work (he submitted his mathematics thesis in 1878) or his work on mining. In March 1879, Poincaré received his degree of "ordinary engineer" from the Ecole des Mines, arriving on 3 April for his new position at Vesoul, where his inspections began the next day and continued intensively over the following months. On 4 June 1879, he reported that Saint-Charles had just about exhausted its pits—"veins both poor and irregular." On 25 September at the pits of Sainte-Pauline, he focused on aeration, removal of gas, and sources of water—just the sort of engineering tasks that Ecole des Mines had emphasized. A month later, on 27 October, Poincaré arrived at the pits of Saint-Joseph to inspect the smelting works. His last mining visit took place on 29 November 1879.

But one stop on the itinerary, at Magny, was anything but routine. In the evening of 31 August 1879, at 6:00 P.M., twenty-two men descended for their shiftwork into the coal pits. About 3:45 A.M. an explosion rocked the mine, instantly blowing out the miners' lamps. Two miners in the cage were violently shaken, two were knocked into the sump (which, luckily, was covered with a board about five feet down). These four survivors staggered to the earth's surface. Master miner Juif, who had been off duty near the pits, immediately led the men back into the mine, where they discovered a pile of clothes burning without flames like a piece of smoldering punk. Juif headed straight for them, extinguishing the material before it could ignite the wooden retaining structures, the coal, or, worst of all, another catastrophic gas explosion. Following cries, they discovered Eugène Jeanroy, a sixteen-year old, who died of his wounds the next day. Every other miner the team found during the search was already dead, some of appalling burns.

Poincaré entered the mine almost immediately after the explosion, in the midst of the rescue operation, despite the risk of a secondary detonation. As the assigned mining engineer, it was his job to sort out what had caused the disaster. Searching for that culpable first spark, he turned first to the lamps. Designed by Humphrey Davy in 1815, these "safety lamps" surrounded an illuminating flame with a closely woven wire mesh that passed light and let air in but kept the flame from escaping. A punctured lamp in a methane-filled mine is a disaster in waiting. Numbers 414 and 417 had belonged to Victor Félix and Emile Doucey; they were never recovered. Lamp 18 was totally destroyed by a cave-in—its mesh and glass no longer anywhere to be found, the rods twisted and broken, and the top entirely detached from the base. His attention, Poincaré wrote in his investigation report, was particularly drawn to lamp 476: it had no glass and two ruptures. The first tear, long and wide, seemed to have come from an interior pressure. The second, by contrast, was rectangular and clearly came from the outside; in fact the puncture was a dead ringer, according to all the workmen, for a strike from a standard-issue miner's pickax. That particular lamp had been signed out to Auguste Pautot, a thirty-three-year-old miner, but it was not found next to his body. Instead, Poincaré observed, lamp 476 still hung from a timber support a few inches from the ground and in the immediate proximity of workman Emile Perroz's corpse. Poincaré and the rescuers found Perroz's own lamp intact, elsewhere.

Throughout his report Poincaré mixed a factual with a more personal tone, one not yet hardened by years of accident inquiries. He recommended the chief miner for compensation in virtue of his bravery, and concluded the medical section with the mournful addendum that he hoped the miners had died instantly, sparing them a long agony. Poincaré's conclusion listed not only the victims but the nine women and thirty-five young children they left behind: "Even the generous efforts of the company will perhaps be insufficient to relieve so much misery."[15]

Exploring the causes of the accident, Poincaré's language turned analytic, advancing hypotheses and counterhypotheses, facing them one by one against the evidence. He acknowledged, for example, the commonplace that miners upstream of an explosion in the air supply are usually burned, while those downstream are asphyxiated. In the Magny disaster, all the deceased suffered burns, so it was logical to think that the explosion had occurred down the airstream from the last man, Doucey. Cave-ins seemed to reinforce this deduction and to suggest that the gas explosion occurred in the half-moon (see figure 2.2). Against this seemingly plausible notion stood another hypothesis that "equally well accounts for the facts." In particular, Poincaré considered whether the explosion might have occurred near the timber peg immediately adjacent to the point where lamp 476 stood:

> So here we are in the presence of two hypotheses up until this time equally plausible: explosion in the half-moon, explosion at the top of the lift. Without the Doucey lamp one cannot prove directly that it had not caused the initial ignition of the gas. But diverse considerations nonetheless support the idea that the primary accident took place in the work space of Perroz.[16]

Since Perroz was a coal loader and therefore had no pick, Poincaré reasoned that it must have been Pautot who had accidentally punctured the lamp with his axe, and then inadvertently exchanged lamps with Perroz. Sometime after that switch, the punctured lamp 476 had lit the atmospheric methane, initiating the conflagration, and setting off a secondary explosion at the point where the incompletely burnt gas encountered the principal flow of air.

Step by step, Poincaré reasoned on through the investigation, eliminating other possible sources of gas one by one: Some sources were outside the flow of air, other veins of coal too old to be degassing. Clinching the argument, he added, was the fact that *any* source of gas from a distant source would have risen to the top of the tunnel and not, therefore, been in contact with the low-hanging lamp 476. Any slow venting of gas would have asphyxiated Perroz who, according to the medical personnel, had died standing. No, Poincaré concluded, the gas must have entered suddenly, probably from a natural gas vent just a single step from lamp 476. When the gas hit the punctured lamp, the miners were doomed.[17]

Poincaré worked the investigation for months. He returned to the mine on 29 November 1879 to follow up on the mechanics of aeration, conducting tests, making measurements of flow, and determining the relative air pressure at different points in the shafts. He submitted his report at Vesoul on the first of December 1879. That same day the ministry of public instruction advised him that he would be given a junior lectureship in the faculty of sciences at Caen. But neither mathematics nor any of his other pursuits ever completely diverted Poincaré from his interest in mining. At the end of 1879 he still hoped to continue his mining engineer career simultaneously with his mathematical one. In fact, he never left the

Figure 2.2 Magny Explosion Map. *Poincaré drew this map to trace the flow of air through the mineshaft at Magny, during his days as a mining safety engineer. On the basis of his investigation, he concluded that the fatal explosion of 31 August 1879 had been caused by a miner's inadvertent puncture of his Davy "safety" lamp rather than because of an explosion in the "half moon" shaft located at the top of the map.* Source: Roy and Dugas, "Henri Poincaré" (1954), p. 13.

Corps des Mines: named chief engineer in 1893, Poincaré became inspector general on 16 June 1910. Just before his death in 1912, he published an article titled "Les Mines" in a book he and a few colleagues produced to join the cultural with the technological and scientific. Poincaré's entry is preceded by a small picture of a Davy lamp without comment, but no doubt a sign for him of the still-smoldering pits of Magny thirty-five years earlier. "One spark," Poincaré wrote later in the piece, "is enough to ignite . . . ; well, I refuse to describe the horrors that follow."[18]

Chaos

Even while sketching and improving the machinery of mining—lamps, lifts, and pit ventilators—Poincaré wrestled with mathematical problems that were, at the same time, physical ones. Those problems converged on what had become *the* great challenge of celestial mechanics: the three-body problem. It is easy enough to state. The motion of a single body is given by Newton's injunction that a body in motion tends to stay in motion. The motion of two bodies, attracted to one another by Newtonian gravity, could also be solved. With the simplifying assumption that the planets were only attracted by the sun (and not by each other), it was a straightforward exercise for Newton and his successors to calculate the precise trajectory of these bodies around the sun. But for a system of three or more mutually attracting objects, such as the sun, the moon, and the earth, the situation was far more difficult. Eighteen interrelated equations had to be satisfied to solve the problem. If space is measured by three axes x, y, and z, then a full description of the motion of the orbs would require the positions x, y, z at each moment in time for each of the three heavenly bodies (that makes nine equations), along with the momentum of each in each direction (another nine). By choosing the right coordinates, these eighteen equations could be reduced to twelve.

For many mathematicians of the mid-nineteenth century, the trend of their discipline was toward an ever-more-rigorous formulation—precise definitions, proofs designed to annihilate the smallest shred of doubt. Such a passion for airtight logical proofs was not what drove the curriculum at Polytechnique, and it never formed one of Poincaré's abiding concerns. Nor was he after better methods to solve equations, though in astronomy such studies could increase the accuracy of predictions about where a planet might be found. Charts of the ephemerides (as the juxtapositions of heavenly objects were called) were very useful to a ship's navigator. Finding such numbers that were both scientifically and practically important might have been just the sort of thing that a Polytechnician like Poincaré wanted. But continuing in the usual way was not an option: he could show that the usual approximation methods for finding the ephemerides gave dramatically wrong predictions of where the planets would be. Never having been drawn to the pure mathematician's fixation on rigor and now convinced of the futility of the applied astronomer's traditional death grip on numerical methods, he needed a dramatically new approach.

Poincaré found that new entry into celestial mechanics though diagrams: he focused on what he called the *qualitative* featues of differential equations. A differential equation tells how a system of things—points, planets, or water—changes from one moment to an infinitesimally later moment. By itself this is not much use in making predictions: knowing where a planet will be an instant from now will not help a ship's navigator. For a useful longer-range prediction, the astronomer had to add up many infinitesimal changes to calculate, for example, where Mars would be next June. To many mechanicians, solving such problems meant doing this adding up (integrating) and putting the final result in a simple, recognizable form. This was not Poincaré's goal.

More or less from the time Poincaré left the mine pits, he aimed to attack differential equations in his own way. Instead of following one water droplet down a stream, so to speak, he wanted to characterize the flow pattern of *all* the drops making up the surface of the water. He was after the general pattern of flow to extract the features of the system as a whole. For example, how many vortices formed—six, two, none? Such an approach would neither provide an elegant formula to yield the velocity of a particular droplet so many centimeters downstream, nor offer a new numerical scheme to approximate the position of Mars next 12 April. Instead, Poincaré sought a *picture* that would capture the physiognomy of the equation and the physical system it represented. Under what conditions would an asteroid or planet fly off into space? Career into the sun? Of course such studies were abstract, mathematical; but they were, at the same time, also concrete. Poincaré wanted, above all, to grasp curves and their qualitative behavior—later could come the details of formulas, numerical predictions, or the last degree of rigor.[19]

By staking his reasoning to the fulness of the visual-geometrical approach rather than the icy edge of algebra, Poincaré was returning, now with far greater sophistication, to the mathematical ambitions of a much older Polytechnique. Not for him the great abstract formulas of Euler, Laplace, or Lagrange. It was Lagrange, after all, who so distrusted geometry that he swore his great work on analytic mechanics would stand on algebra alone, never would it rest on geometrical constructions. Not a mechanical analogy, not a single diagram would sully its pages.

By contrast, Poincaré worked precisely in such geometrical ways, and mechanical analogies were near to hand. Already in 1881, when focusing differential equations that were linked to the three-body problem, he highlighted his qualitative, intuitive ambition:

> Could one not ask whether one of the bodies will always remain
> in a certain region of the heavens, or if it could just as well travel
> further and further away forever; whether the distance between two

bodies will grow or diminish in the infinite future, or if it instead remains bracketed between certain limits forever? Could one not ask a thousand questions of this kind which would all be solved once one understood how to construct *qualitatively* the trajectories of the three bodies?[20]

Here, in the joint realm of the geometrical-visualizable and the physical, were concerns Poincaré returned to time after time, just as he came back throughout his career to questions in the mathematics of differential equations and the physics of the three-body problem. As he put it in 1885 (about one of his most important pieces of mathematics), one simply "cannot read . . . parts of this memoir without being struck by the resemblance between the various questions which are treated there and the great astronomical problem of the stability of the solar system."[21] Mechanics — machines — always lay near. It was the factory stamp.

Poincaré pushed ahead with his qualitative program for understanding the behavior of differential equations, with success sufficient to catch the attention of the world's leading mathematicians. When, in mid-1885, *Nature* printed an announcement for a mathematical competition honoring the 60th birthday of Oscar II, the King of Sweden, Poincaré was a leading contender to win. Gösta Mittag-Leffler, a well-known mathematician and editor of *Acta Mathematica*, had responsibility for gathering the committee of judges. First he recruited Charles Hermite, one of Poincaré's Polytechnique teachers, and then Hermite's own professor, the formidable German mathematician Karl Weierstrass, whose mathematical life had long stood for relentless logical rigor. (Mittag-Leffler was himself on friendly terms with Poincaré.) Entries were due 1 June 1888: the very first question was on the three-body problem.[22]

Between the announcement of the prize and the submission date, the French Academy of Sciences elected Poincaré to its ranks. This was a signal honor. It meant that, as of 31 January 1887, at the age of thirty-two, he was a fixture of the French scientific elite. Membership in the Academy meant he could be (and frequently was) called to a wide range of administrative functions, from the Bureau of Longitude to interministerial commissions ranging across legal, military, and scientific domains. Adjusting easily to this new public role, Poincaré began writing for a broader audience. Of the nearly 100 nontechnical articles and books he penned for newspapers, journals, and scientific reviews, you can count on one hand those written before his appointment to the Academy of Sciences.

Throughout the whole of his career, visual, intuitive methods served as Poincaré's pole star. He had used non-Euclidean geometry as a way of cracking problems at the very beginning of his work. And now, using visual (topological) techniques that he had honed during a decade considering differential equations, Poincaré entered the prize competition under his Latin banner, "For the stars do not cross their prescribed limits." In that adage, much was conveyed. Technically, Poincaré was aiming to set bounds on the motions caused by the mutual attraction of the planets, reaffirming the stability of the solar system. But beyond the mathematics, Poincaré's motto reflected his deep faith in the basically stable nature of the world around him. Despite the excellence of several runner-up contributions, Poincaré won the competition hands down; offering not only results but also a host of new methods that placed his achievement at the pinnacle of mathematics. All was, or seemed, right with the world.

After submitting the prize essay (perhaps thinking of the role that his own qualitative, visually oriented work had played in showing the stability of the Solar System), Poincaré stepped back to survey the role of logic and intuition in mathematical science and pedagogy. Looking over older mathematical books, he said, contemporary mathematicians saw work punctuated with lapses in rigor. Many of the older concepts—point, line, surface, space—now seemed absurdly vague. The proofs of "our fathers" looked like frail structures unable to support their own weight. Poincaré acknowledged that mathematicians now knew, as their fathers did not, that there are crowds of bizarre functions that "seem to struggle to resemble as little as possible the honest functions that have some useful purpose." These new functions might be continuous but they are constructed in such a peculiar way that one cannot even define their slope. Worse yet, Poincaré lamented, these strange functions seem to be in the majority. Simple laws seem to be nothing but particular cases. There was a time when new functions were invented to serve practical goals; now we mathematicians invent them to show off the faulty reasoning of our fathers. If we were to follow a strictly logical path, we would familiarize beginners, from their start in mathematics, with this "teratological museum" of freakish new functions.

But this teratological path was not one Poincaré counseled his readers to follow—whether students or pure mathematicians. In mathematical education, he argued, intuition should not be counted least among the faculties of mind to be cultivated. However important logic was, it was by way of intuition "that the mathematical world remains in contact with the real world; and even though pure mathematics could do without it, it is always necessary to come back to intuition to bridge the abyss [that] separates symbol from reality." Practicians *always* needed intuition, and for every pure geometer there were a hundred in the trenches. Even the pure mathematician, however, depended on intuition. Logic could provide demonstrations and criticisms, but intuition was the key to creating new theorems, new mathematics. Poincaré was blunt: a mathematician without intuition was like a writer locked in a cell with nothing but grammar. His view was therefore that instruction (he referred explicitly to Polytechnique where he was by then teaching) should emphasize the intuitive and abandon these formal, unintuitive functions that only served to haunt the mathematical legacy of our mathematical ancestors.[23]

This plea for mathematical intuition appeared in print as Poincaré's prize paper was about to be published. But in July 1889, Edvard Phragmén, a twenty-six-year-old Swedish mathematician working as an editor of *Acta* under Mittag-Leffler noticed some problems with the prize proof. He relayed them to Mittag-Leffler, who cheerfully told Poincaré on 16 July that, with a single exception, "one could make them disappear almost immediately."[24] Poincaré soon understood that the exception was not easily repaired; this was neither a typographical error nor a simple gap easily plugged with a few lines of more careful mathematics.[25] Something was deeply wrong with his work. Not only the prize, but his, the journal's, and the judges' reputations were at stake; Poincaré had to figure out what was wrong.

Here was the issue. As in his studies of differential equations, Poincaré had considered three bodies: a little asteroid hurtling around the orbiting system of Jupiter and the Sun. What could the orbit of the asteroid do? One particularly simple behavior would be to return every time to the same spot with the same velocity: simple periodic motion. To represent such repetitive orbits in a dramatically simpler way, Poincaré proposed a startling idea: Don't think about the trajectory itself. Instead, Poincaré realized that he could examine the situation once each time the asteroid came around— creating a stroboscopic picture, so to speak, that has come to be known as a "Poincaré map." Properly speaking, his map plotted the asteroid's momentum and its position each time around, but we can capture the idea by picturing the map as a vast sheet of paper, much bigger than any planet, spread in space perpendicular to the asteroid's wandering. Every time the asteroid comes back, picture it punching a hole F in this cosmic sheet. In a simple periodic orbit, the asteroid would punch through the sheet and then travel through that same hole over and over again for all eternity. That hole is termed a fixed point. More generally, Poincaré's idea was to study the patterns punctured in the two-dimensional sheet by the asteroid rather than its full orbit in space.

If the asteroid began its orbit at a different position and speed, however, it need not travel in such a simple, repetitive way. For example, imagine that another, identical asteroid crossed the sheet near, but not through, the hole *F*. One possibility is that the punch-throughs would, orbit by orbit, drill a succession of holes through the sheet that approached *F*, and, in the fulness of eternity, reach it. Imagine a curve *S* drawn through successive holes. This curved axis *S* through *F* is called *stable*, if an asteroid starting anywhere on the line (that is, whose punch-throughs begin anywhere on the curve) gradually tends toward an orbit that circulates through hole *F* every time (see map 1 in figure 2.3). Conversely, a curve through *F* is called *unstable* if, in the distant past, an asteroid punching *F* gradually moves away from *F* on the curved axis *U* (see map 2).

It was Poincaré's claim in his prize paper that if an asteroid's punch-throughs fled from a fixed point like *F*, they would eventually settle down toward another fixed point. Such a result would have described an ordered, bounded world, a world fully matching Poincaré's hopeful motto of stars hemmed within their limits. Pushed by Phragmén to study his argument more closely, Poincaré dissected the problem during the fall of 1889, and his confidence in planetary stability began to crumble. Meanwhile, publication of the prize paper went forward.

On Sunday, 1 December 1889, Poincaré confessed to Mittag-Leffler:

> I will not conceal from you the distress this discovery has caused me. In the first place, I do not know if you will still think that the results which remain, that is the existence of periodic solutions, the asymptotic solutions [, and my criticism of earlier methods] deserve the great reward you have given me.
>
> In the second place, much reworking will become necessary and I do not know if you can begin to print the memoir; I have telegraphed Phragmén. In any case, I can do no more than confide my perplexity to a friend as loyal as you. I will write to you at length when I can see things more clearly.[26]

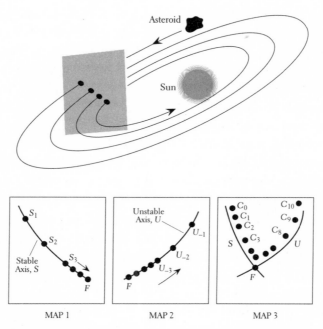

*Figure 2.3 **Poincaré's Map.** Poincaré's "stroboscopic" diagrams tracking the intermittent punctures of a sheet so resembled a cartographic representation that they became universally known as "Poincaré maps," with features commonly referred to as "islands," "straits," and "valleys." Map 1 depicts a stable "axis" S along which successive punctures converge toward the fixed point, F. Map 2 shows an unstable axis U with successive punctures running away from F, and Map 3 portrays the crossing of both an unstable and stable axis at F. In this last case a puncture C_0 near the curved stable axis S is followed by subsequent hits progressing toward F (while remaining not far from S), and then fleeing from F (while remaining alongside the unstable curved axis U).*

Wednesday, 4 December: Mittag-Leffler wrote Poincaré to report how "extremely perplexed" he was to hear from Phragmén of Poincaré's assessment of the situation. "It is not that I doubt that your memoir will be in any case regarded as a work of genius by the majority of geometers and that it will be the departure point for all future efforts in celestial mechanics. Don't therefore think that I regret the prize. . . . But here is the worst of it. Your letter arrived too late and the memoir has already been distributed." Write a letter to me, he continued, in which you explain that, based on your correspondence with Phragmén, you found that the expected stability was not, in fact, proven for all cases, and that you will send me a corrected manuscript. Mittag-Leffler added that a good word for Phragmén would be apt—a chair was open at the university. "It is true that my adversaries, acquired through the success of the *Acta*, will make a scandal out of this, but I will take it with tranquility because it is no embarrassment to be mistaken at the same time as you, and because I am firmly persuaded that you will eventually unravel the most hidden mysteries of this extraordinarily difficult question."[27]

The next day Mittag-Leffler swung into action, reporting to Poincaré that he had telegraphed Berlin and Paris to demand that they not distribute a single copy of the flawed journal. In Paris only Charles Hermite and Camille Jordan had received copies; Karl Weierstrass had one in Berlin. To Jordan, for example, Mittag-Leffler wrote to say that an error had slipped in and had to be fixed—could he please leave his copy to be picked up by a "domestique" who would whisk it away. "Please don't say a word of this lamentable story to anyone," he urged Hermite, "I'll give you all the details tomorrow." Tracking down the other copies one by one, Mittag-Leffler began to hope that all copies were recoverable. "I am very glad," he confided to Poincaré, "that Mr. Kronecker [an equally famous German mathematician and arch enemy of Weierstrass] had not received a copy." But even Mittag-Leffler's allies began to bridle at the campaign. When Weierstrass replied to Mittag-Leffler's sugar-coated letter, his discontent showed: "I confess to you, furthermore, that I take the matter not so lightly as you, Hermite and Poincaré himself." Weierstrass icily noted that in his country, Germany, it was axiomatic that prize essays were printed in the form in which they had been judged. Weierstrass added that the stability question was not peripheral to Poincaré's essay; rather, as Weierstrass had indicated in a report that was to have served as an introduction to Poincaré's essay, it was central. What, Weierstrass demanded, remained in Poincaré's paper of its entire positive program?[28]

Poincaré rewrote the paper. What remained (or rather what he created to plug the gap), was something altogether outside the range of possible motions he — or anyone else — had ever considered. Chaos, not stability, reigned in this new universe. Here is what happened. Suppose, along with Poincaré, that a line of stability and line of instability cross at the fixed point F. (This is not hard to imagine. Consider a saddle: a marble sent down straight from the pommel along the direction of the horse's spine will oscillate back and forth until it settles in the middle of the saddle: a stable point. But a marble propelled down to the right or left will fall off, never to return: an unstable point.) Now, suppose our asteroid begins near but not on the stable axis; it will gradually move toward F until it gets quite near, at which point it will fall under the sway of the unstable axis and begin to wander away from F. This is depicted in figure 2.3, map 3, as the series of punch-throughs, $C_0, C_1, C_2, \ldots C_7, C_8, \ldots$ So far, Poincaré had no problem.[29]

But suppose the stable and unstable axes cross somewhere else, for example at H, which Poincaré dubbed a *homoclinic* point. H will then be, by assumption, *both* a point on the stable axis S (driving the subsequent punch-throughs of the asteroid toward F) and a point on the unstable axis U (so an asteroid beginning near F will, slowly at first and then more quickly, puncture its consequents [successive holes] in sequence away from F on U through H). Now suppose that an asteroid flies through H; since it must remain on S, it beats its way, in each subsequent crash through the sheet (H_1, H_2, H_3, etc.) along S toward F. But since any point on the unstable axis always remains on U, since H is also on U, all the consequents of H also have to be on U. So the extended U-axis must hit every consequent of H that we just plotted: H_1, H_2, H_3, and so on. One way this might happen is shown in figure 2.4a.

Now consider a different asteroid, C, near H but on U (figure 2.4b). Because H is on S, C (being near S) will start to move toward

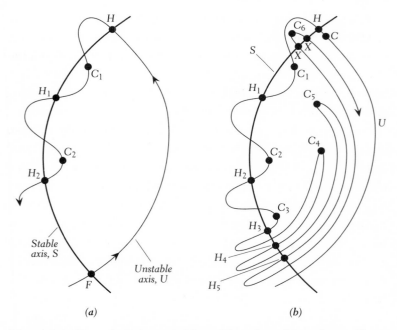

(a)   (b)

Figure 2.4 Chaos in Poincaré's Map. *When the stable and unstable axes cross, complexity can follow. Indeed, it was precisely the possibility of this chaos-inducing crossing that Poincaré left unresolved in his first submission of the prize essay. As described in the text, this leads to the enormously complex extension of the unstable axis that begins in* (a) *and then is developed more fully in* (b). *Even the behavior of* (a) *is just the beginning of the complexity that would follow as the new crossing points of S and U (at the points marked X) are taken into account. Understandably, Poincaré despaired of ever being able to draw the "latticework" that a more complete representation would have to show.*

F. This is just what we saw in figure 2.4a. At the same time, since C starts on U, its consequents (C, C_1, C_2, C_3 and so on) have to stay on U as U is extended. So the series C, C_1, C_2 moves toward F. But eventually, as the consequents of C approach F, the subsequent C's begin to beat a retreat along the unstable axis U as in figure 2.3, map 3. Since U crosses S at H, the C's, in their retreat along U, also eventually cross S (here illustrated by C_6). The U-axis, after crossing S at H_3, has to zoom all the way back to C_4 to hit it, and then the U-axis has to head back once again toward F to intersect S at H_4. Notice that because the extended U-axis to and from C_6 has hit the S-axis (at the two points labeled X) we have two *new* homoclinic points and the map becomes vastly more complicated yet.

As a result of all this complexity, far from settling into bounded behavior, the unstable axis, and therefore any asteroids on or near it, will wander all over the Poincaré map, creating a motion so extraordinarily complex that Poincaré himself proved unable to depict it. When he eventually expanded the prize essay for publication in his *New Methods of Celestial Mechanics*, he struggled to describe the figures that resulted:

> When we try to represent the figure formed by these two curves and their infinitely many intersections, each corresponding to a doubly asymptotic solution, these intersections form a type of trellis, tissue, or grid with infinitely fine mesh. Neither of the two curves must ever cut across itself again, but it must bend back upon itself in a very complex manner in order to cut across all of the meshes in the grid an infinite number of times.

"I shall not even try to draw it," he added, yet "nothing is more suitable for providing us with an idea of the complex nature of the three-body problem."[30]

A hundred years after the vexed publication of this prize essay, Poincaré's exploration of chaos was all the rage, celebrated as the dawn of a new science, a revolutionary advance over the simple predictions of classical science. Some late-twentieth-century physicists, philosophers, and cultural theorists hailed the sciences of complexity (as they came to be called) as a form of "postmodern physics," while powerful computers spewed Poincaré maps minutely illustrating what their inventor had despaired of seeing on paper. Some of these maps revealed new physical phenomena; others graced art galleries.[31] But in 1890, Poincaré proposed no revolution in the nature of science. Confronting a potentially damaging scandal over the prize, he patched up the gap in his argumentation and, in exploring the new dynamics, found what he had neither sought nor desired—a crack in the stability of the Universe.

Far from waving a radical banner, Poincaré showed that, although the number of such chaotic orbits was infinite, the likelihood that an asteroid would find itself in an unstable regime was insignificant as measured against the likelihood that its orbit would be in a stable one. Having lost an absolute stability, Poincaré had to settle for a probabilistic one: "One could say," he wrote about two years after the revised paper, that "the [unstable orbits] were the exception and the [stable ones] the rule." Instead of trumpeting a break from stability, Poincaré stressed the power of the new qualitative methods to explore classical celestial dynamics: "[T]he true aim of celestial mechanics is not to calculate the ephemerides, because for this purpose we could be satisfied with a short-term forecast, but to ascertain whether Newton's law is sufficient to explain all the phenomena."[32] For Poincaré, the true trial of Newton's physics would come by probing its qualitative features: "From this point of view [of ascertaining the sufficiency of Newton's law], the implicit relations which I have just spoken of can serve just as well as explicit formulas."[33] At stake for Poincaré were the basic, underlying relations of things, not the formulae and positions that astronomers busied themselves calculating to ever-higher decimal places.

Convention

In Poincaré's attention to structures rather than things, there is a striking analogy to his use of non-Euclidean geometry. For many of Poincaré's contemporaries, the non-Euclidean geometries explored throughout the nineteenth century marked an arresting break. Euclidean geometry had, for centuries, practically defined proper reasoning from sure starting points to inevitable conclusions. Since the eighteenth century, philosophers reading (or, perhaps, misreading) Kant, had enshrined Euclidean geometry as knowledge built into the very fabric of the mind. Some scientists and philosophers saw non-Euclidean geometry as a radical shift in the very definition of knowledge, a sign of hopeful modernity that broke completely with intuition; others feared the dreadful loss of certainty. Poincaré maintained a much more pragmatic stance. On the one hand, he insisted in 1891 that if the Euclidean axioms could be known before any experience, we humans could not so easily imagine others. On the other hand, the axioms of Euclidean geometry could not be merely experimental results. If they were, we would be constantly revising them. Since there are no perfectly rigid bodies that we must use to instantiate straight lines, we would soon "discover" that geometry was simply false: we would find, for example, triangles with angles that did not sum *exactly* to 180 degrees. Poincaré insisted that our choice of geometry is *guided* by experimental facts but is ultimately open to choice, subject to our need for simplicity.

One of Poincaré's more broadly directed pieces dealt with simplicity, convenience, and the hypotheses of geometry. What are the hypotheses? In what sense are they true? For Poincaré geometry was nothing other than a *group*, that is, a set of objects along with an operation that has certain properties. One property of the operation is that it should be reversible. The integers (. . . -3, -2, -1, 0, 1, 2, 3, . . .) have this property under addition and subtraction; adding a whole number can always be reversed by subtracting the same number. The group should also have among its possible actions that of identity, the operation that leaves a given object unchanged; here, adding zero does just that. And combining operations should lead to an element still in the group: adding 5 and adding 8 gives the same effect as adding 13. We learn which groups are useful to us by our encounter with the world but—as Poincaré often insisted in his philosophical writings—the idea of the group itself was a tool with which we were born. Not surprisingly, one particularly interesting group for us humans was the movement of rigid bodies in space. Poincaré argued that we have picked ordinary Euclidean geometry from among many other choices because the group underlying it corresponds, in a simple way, to the group of rigid-body motions in space—solid objects move in our real world. Could one have chosen otherwise? Of course, Poincaré says: we have only chosen the most convenient geometry. Does that mean that the other geometries are false? Not at all. No more (according to Poincaré) could one promote Cartesian coordinates (measuring a point's position by the usual x and y axes) as true and derogate polar coordinates (measuring a point's position by its distance from the origin and the angle of that radius line) as false. Once again, Poincaré emphasized that there are free choices in representing the world, choices fixed not by something completely exterior, but rather fixed by the simplicity and convenience of our knowledge. "I won't insist any further; because the goal of this work is not the development of these truths which are starting to become banal."[34]

In fact, Poincaré was so committed to the role of conventional choice that he argued that the selection of a geometry was like the choice between French and German: Poincaré contended that one can choose this or that language or idiom to express the same thoughts. Imagine intelligent ants who lived on the surface of a saddle, and defined straight lines as the shortest distance between two points. Ant Mathematician says: "The sum of the angles of a triangle is less than two right angles." We (from our human perspective) would describe the same situation in "Euclidean" rather differently, because we would see the ants' triangles as shapes with curved sides. Human Mathematician might say: "If a curvilinear triangle has for its sides arcs of circles that, if produced, would cut orthogonally the fundamental plane, the sum of the angles of the curvilinear triangle will be less than two right angles." Both statements capture the same situation but in different idioms. So there is no contradiction. Such a correspondence shows that the theorems of ant-saddle geometry are no less consistent than our ordinary geometry; they might even be useful: a saddlelike "geometry being susceptible of a concrete interpretation, ceases to be a useless logical exercise, and may be applied." *That* is the punchline: Various geometries are simply different ways of presenting relations among things; which we use depends on convenience.[35]

All through Poincaré's mathematics, his philosophy—and as we will see, his physics—lay this productive theme: find the group structure of what can be altered, and choose the most convenient representation. Yet while making those free decisions never lose sight of the fixed points, the invariants—those bits of the world left unchanged by our choices.

WHERE, THEN, WAS Poincaré by 1892? The rising star of French mathematics, he had made his name through work that made dramatic use of non-Euclidean geometry, advances in qualitative approaches to differential equations, and a stunning new approach to the three-body problem. In philosophy, geometry, and dynamics, he had begun to work out a vision of knowledge that had two simultaneous goals. De-emphasizing particulars, he found the free choice available in formulating the problem. The choice of a coordinate system was not given by nature but made by us for our own convenience. Similarly, particular approximation schemes were to be selected by us for specific purposes; even the choice of a specific geometry had no absolute importance. Poincaré's view: use Euclidean geometry when it is useful, when it pays to employ a non-Euclidean geometry, use that. In differential equations or in the physical systems they represented, there were always many ways of choosing variables—to describe, for instance, the flow lines of water down a stream. What was significant were the underlying relations that remained unchanged even after such changes in description: the vortices in a flow of water, the knots, saddle points, or spiral endpoints of geometrical lines. Similarly, the length of a line remains fixed when we rotate coordinates.

These two aspects of Poincaré's work—the variable and the fixed—emerged together and can only be understood together. He says, in different ways over many years: Manipulate the flexible aspects of knowledge as tools; choose the form that makes the problem at hand simple. Then seize those relations that stand fast despite the choices made. Those fixed relations stand for knowledge that endures. Together the variant and invariant make scientific progress possible.

For a hundred years, scholars have struggled to get at the root of Poincaré's conventionalism. Some, with good reason, have stressed the role of geometry. As Poincaré emphasized over and again, however, mathematical statements can be made in the language of non-Euclidean just as well as in Euclidean geometry. Other scholars have scrutinized the works of earlier figures in geometry—such as Felix Klein, the great German geometer who propagandized so forcefully for different kinds of geometry. Still others have gone back to Sophus Lie for the root of Poincaré's picture of free choice among geometries. After all, Poincaré explicitly cited Lie as a mathematical forebear, and Lie was quite clear on the arbitrariness of many choices made by the mathematician. For example, Lie said that Descartes identifies the variables x and y by a point on the plane. But (according to Lie) Descartes could, "with equal validity" have chosen to symbolize x and y by a line, and develop geometry from that asssumption. Moreover, Descartes defined x and y according to a specific coordinate system—the x and y refer to their distances from the x- and y-axes. There too is a certain free choice: "[P]rogress made by geometry in the 19th century," wrote Lie, "has been made possible largely because this two-fold arbitrariness . . . has been clearly recognized as such." Lie was here contending that mathematical progress followed from the recognition that there were always many ways to represent mathematical concepts. Choosing the particular representation of mathematics, choosing this or that geometry is, Lie argued, a matter of "advantage and convenience." It was, as one scholar has persuasively argued, "one of the grounds of Poincaré's commitment to a view of geometry that held our choice of geometries to be one of open choice, grounded by convenience."[36] No doubt Poincaré's emphasis on free choice in geometry can be hunted back to the German polymath Hermann Helmholtz, who struggled to disentangle factual meaning from definitions in geometry, always emphasizing the central role of mobile, rigid objects in fixing our concepts of space. It may be too that Poincaré's ideas on mathematical conventionalism should be pursued back to the myriad geometries of Bernhard Riemann, or for that matter to the more recent work of Poincaré's teacher, Charles Hermite.[37]

Propelling this sense of a freedom of choice was a pedagogical *conventionalism* (if you will) to be found in the stridently agnostic instructional style of the Ecole Polytechnique. That abstention from absolute commitment to any particular theory was a prominent feature of Alfred Cornu's courses, which Poincaré had attended as a student. Alternative theories each had advantages and disadvantages; all were constrained only by the much-emphasized experimental fixed points. For Poincaré, the invariants of physics (which provided objective knowledge) were the fixed relations between

experiments, relations that survived the ever-changing flux of theories. Recall that the same free choice among theories had informed
Polytechnique's even-handed hiring—some representative scientists
for atomism, some against. Such an abstention from absolute theoretical commitment characterized Poincaré's own courses; lectur-
ing on electricity and optics in 1888, 1890, and 1899, for example,
he gave each major theory its moment in the sun, displaying its
virtues and vices for the students to judge. A rhyzomatic "root" of
conventionalism here too.

Finally, in Poincaré's exchanges with his brother-in-law's (Emile
Boutroux's) philosophical circle, Poincaré would have found conventionalism in a philosophical register. This loose affiliation of
philosophers and philosophically minded scholars offered Poincaré,
early on, a more reflective view of the mathematical sciences. Raw
empiricism was avoided as woefully inadequate to account for the
generality and extent of scientific knowledge. Pure idealism (reducing reality to mental life) could not explain the concordance of
ideas with the world. Drawing strongly on the Kant revival under-
way in Germany, Boutroux and his circle rejected both the
extremes of idealism and empiricism. Taking science and the
humanities to be inextricably bound, these philosophers saw both
structured by an active role for the mind and a suspicion toward the
purely metaphysical. In his encounters with Auguste Calinon's
work on the philosophical foundations of physics, Poincaré walked
this philosophical middle line straight toward the problem of
simultaneity.

Geometry, topology, pedagogy, philosophy—each of these ways
of parsing Poincaré's world tells us something about the ways in
which scientific "free choice" made sense to him. Intriguingly,
around 1890, Poincaré began regularly to call "free choice" by a
new name, insisting (as he had in 1887) that the geometrical axioms
are neither experimental facts nor (as some Kantians would have it)

printed in advance on our human minds. In a curt, insistent sentence printed in 1891, he lay down a new formulation of his view of geometric axioms: *"They are conventions."*

> Is Euclidean geometry true? It has no meaning. We might as well ask if the metric system is true, and if the old weights and measures are false; if Cartesian co-ordinates are true and polar co-ordinates false. One geometry cannot be more true than another; it can only be more convenient. Now, Euclidean geometry is, and will remain, the most convenient.[38]

Here the status of the axioms of geometry have been explicitly likened to terms in a language that can be freely chosen and (in this quotation) also to the freedom the mathematician or physicist always has to choose a coordinate system. The new element is that Poincaré depicts the choice of Euclidean or non-Euclidean axioms not just as a choice among groups, but as a choice between the arbitrary system of meters and kilograms and the arbitrary system of feet and pounds.

To appreciate this aspect of Poincaré's use of "convention," we must recognize that his reference to weights and measures contains the trace of an entire world of conventions. At the same time, as we will see, Poincaré's concern with meters and seconds cannot be considered an external "influence," one that determined his scientific and philosophical work the way a hidden magnet rearranges the iron filings above it. "Roots" and "influences" are terms too weak and external to capture Poincaré's thoroughgoing engagement with the practice of setting planetwide conventions.

The world of decimalized, conventionalized time and space was anything but abstract to Poincaré. He flourished in (and contributed to) the Parisian, indeed worldwide skein of wires, meetings, and international treaties. Here was, after all, the consummate Polytechnician who moved as easily in the depths of a coal mine as in the far reaches of astronomical stability. But to make visible the mechanisms of clocks, rods, and wires—above all, to grasp the production of the late-nineteenth-century conventional understanding of simultaneity—it is necessary to pull back to a wider perspective. We need to move back and forth, in and out, between the details of work in philosophy, mathematics, and physics and the larger-scale social and technological conventionalization of time and space in which Poincaré took part.

In to the precision swing of master clock pendulums, *out* to the undersea telegraph cables crisscrossing the oceans. *In* to follow the minutae of individual train schedulers, jewelers, and astronomers; then back *out* to the legal recalibration of national and world-covering time zones. In this process of scrutiny, historical light necessarily plays off the very different scales utilized by technological, scientific, and philosophical activity. Between 1870 and 1910, conventions of space and time scintillated with a critical opalescence.

Chapter 3

THE ELECTRIC WORLDMAP

Standards of Space and Time

PARIS, HÔTEL DES AFFAIRES étrangères, 20 May 1875, 2:00 P.M. Represented by their decorated plenipotentiaries, seventeen names will be put to a treaty, their resplendent titles marching across the page: "His Majesty, the Emperor of Germany," "His Majesty, the Emperor of Austro-Hungary," "His Excellency, the President of the United States of America," "His Excellency, the President of the French Republic," "His Majesty, the Emperor of All Russia. . . ." We are at the solemn signing of the Convention of the Meter. After years of negotiation, the High Contracting Parties now called into existence an international bureau of weights and measures. The new prototypes of the meter and kilogram it was charged with certifying would supplant the myriad of competing national measures, establish the relation between these gauges and all others, and compare results with the standards used to map the earth.

Here, in the *convention*, diplomacy met science. When Duke Louis Decazes, the French minister of foreign affairs, sent out invitations to other countries back in 1869 for a diplomatic conference on this issue, he invited politicians, but also leading scientists like the German astronomer Wilhelm Förster, who was director both of the German Bureau of Weights and Measures and the Berlin Observatory. By March 1875 the committee had come far enough for Decazes to gently retire the scientific domain, in which the assembled held only a "relative competence," in order to focus on "questions of a political and conventional order (*ordre convention-nel*)," where they had "absolute competence": their conclusions would form the basis for binding international law. Conventions joining scientific and legal technologies had been concluded before—in 1865, for example, governing telegraphy. Indeed dozens of conventions had aimed to smooth collisions between countries in trade, post, and colonization. Now, in the vital domain of the metric system, even more than the telegraph accord, the delegates had produced an "international contract" as dear to scientists as it was to industrialists and politicians: a legal document that would rule from the spotless precision of the physics laboratory to the smoke and steam of the factory.[1]

If Decazes spoke for diplomacy, Jean Baptiste André Dumas, organic chemist and, since 1868, perpetual secretary of the French Academy of Sciences, spoke for French scientific enthusiasm. As head of the special (scientific) commission on the meter, Dumas had been responsible for the recommendations that now stood before his colleagues. Partially summarizing, partially lobbying, Dumas stood before his fellow delegates to advocate a permanent bureau in Paris vested with the authority to set, maintain, and distribute international standards. Above all, Dumas wanted to justify the universal meter as a standard for industry, for science, for France, for the world. As he saw it, anyone who had set foot in London's 1851 Universal Exposition immediately recognized that "chaos" reigned between national systems. Each country's peculiar system of weights and measures made comparison among them impossible without tedious calculation. At the same time, every subsequent exposition had demonstrated that the reach of the metric system was steadily growing. Everywhere people wanted to throw out discordant measures; they yearned to smash intellectual barriers between peoples. For Dumas, indeed for many senior French scientists, the call for international standards would be heard by all "enlightened men." Having embraced the metric system throughout physics and chemistry laboratories, scientists now taught it widely. Factories, builders, telegraphs, and railroads had seized the

meter. Now, Dumas urged, public administration should back the
rational meter.

Dumas: Decimals mattered. For both sides, practical and pure
science, it was the decimal character of the metric system that mattered. Twelve inches in a foot, three feet in a yard—neither plumber
nor physicist could cherish such a hodgepodge. "As for the geodesic
origin of the metric system," that pride of the French Revolution, by
now "it is absolutely without interest for commerce, for industry and
even for science." Upon its adoption in 1799, the meter was supposed to be exactly one ten-millionth of a quarter of the earth's circumference. Dumas assured his listeners that modern proponents of
the metric system made no such claim; the assembled knew perfectly well that the earth's size could not be measured with the precision needed for an international standard. For Dumas, the reason
for adopting the metric system was because it divided lengths into
sensible units of ten. That was what pure scientists wanted and what
pragmatic journeymen demanded. To spread this new rational system a center was needed. It should be "neutral, decimal, international." It should be *ça va sans dire*, in Paris.[2]

Dumas reminded his audience that the metric standards had
become international precisely because revolutionary France
had designed the system to make it so. Long ago, ancient Hebrews
had put their measuring prototypes in the Temple. Romans set their
standard in the Capitol, Christians sequestered theirs in the Church
(which was how Charlemagne's standard kept its original purity).
For eighty years, the Archives had performed this task for France,
preserving the standard meters since revolutionary times. But now
that the high contracting parties had decided to make the meter a
truly international standard, they judged the revolutionary meter
neither strong enough nor sufficiently invariable to serve as the prototype for the world's measures.

Signing the Convention of the Meter started, rather than ended,
the process of distributing the meter. Bureaucrats and scientists lob-

bied, bullied, and negotiated their countries toward putting the scheme into practice. Some of the great experimenters of Europe and the United States contributed to it: Armand Fizeau, who had measured the "dragging" of the ether by water, as well as the American Albert Michelson, who invented the interferometer, an instrument capable of measuring length to within a fraction of the wavelength of visible light. For fourteen years, French engineers and British metallurgists hammered and smelted their way to a tough, durable iridium-platinum alloy.

While a British firm pounded these hard, pure bars into meter sticks with an inflexible "X" cross section, the French concentrated on producing an enormous "universal comparator" (see figure 3.1), that would, by strict procedure, allow a standard length to be reproduced on another bar to within two ten-thousandths of a millimeter. It was painstaking, nerve-wracking work. When the British metal workers delivered their precious bars to the French, the operator at the conservatoire would set both the standard meter and the blank bar on the bridge of the comparator. Peering through a microscope (M), the operator would line up the one-meter mark on the standard. Then the operator would activate a lever, causing a diamond blade to inscribe a fine line precisely at the one-meter point on the blank. Carving subdivisions was just as difficult. The two microscopes would be set, say, ten centimeters apart. The operators would mark that length. Sliding the bar down, they would etch a second ten-centimeter length into the bar, and so on. To prepare the 30 standard bars that the international delegates would take home with them, the operators repeated this operation 13,000 times. The slightest slip with the diamond point meant starting over again with repolishing of the blank.[3]

Finally, on Saturday, 28 September 1889, two years after Poincaré was elected to the Academy of Sciences, eighteen representatives of the contracting parties gathered in Breteuil for the final sanctioning of the meter. The president of the conference can-

Figure 3.1 Universal Comparator. This machine served to rule precise lengths for platinum-iridium copies of the standard meter, **M**. For engineers, physicists, politicians, and philosophers—especially in France—the international success of the standardized unit of length served as a model for what they hoped would be the decimalization and standardization of time. SOURCE: GUILLAUME, "TRAVAUX DU BUREAU INTERNATIONAL DES POIDS ET MESURES" (1890), P. 21.

vassed their votes—unanimous—and then pronounced: "This prototype of the meter will from now forward represent, at the temperature of melting ice, the metric unit of length," while "this prototype [kilogram] will be considered from now on the unit of mass." All standards stood on display in the meeting room: meters sheathed by protective tubes, kilograms nested in triple glass bell jars. According to plan, each delegate ceremoniously picked a ticket from an urn, the number received assigning his country a meter stick, for which he offered a signed receipt.

Suddenly, these carefully scripted proceedings ground to an abrupt halt. The most important act—the deposit of the meter in its underground safe—was possible only with the three keys needed to open the vault. One of those keys would be in the hands of the director of the French Archives, but he was not there. The president suggested they ask for instructions from the French minister of commerce, but the delegates vigorously objected. Swiss astronomer Adolph Hirsch insisted that the conference was international, not French. The conference would not address an ordinary French minister. Out of the question: Hirsch and his colleagues would deal with France only through its minister of foreign affairs. Diplomacy apparently produced the missing key.

Later that afternoon, at 1:30 to be precise, the commission charged with depositing the international prototypes gathered in the lower basement of the Breteuil Observatory. There the delegates certified that the international prototype **M** would from that moment forward be enclosed in a case covered on the interior with velvet, lodged within a hard cylinder of brass, screwed tight, locked, and placed in the vault. Alongside **M** the standard bearers then prepared two "witnesses" for burial (meter sticks, not delegates). These metallic observers would forever testify, by the very conditions of their bodies, to anything that might befall **M**. In the same ceremonial interment, convention delegates sanctioned the kilogram, **K**, elevating and renaming it as the universal standard of mass. It too found its eternal resting place in the underground iron vault in the company of its witnesses. With two keys, and in full view of the delegates, the director of the International Bureau of Weights and Measures locked the case, secured the inner basement door with a third key, and bolted the exterior door with a fourth and a fifth key. At the conclusion of these solemn events, the president of the conference handed these latter keys in separate, sealed envelopes: one to the director of the International Bureau, one to the general guard of the National Archives, and the last to the president of the International Committee. From that time on, all three basement keys would be needed to enter the sanctum sanctorum.[4]

This was a remarkable moment. **M**, the most precisely forged and measured object in history, the most individually specified humanmade thing, had become, by its burial, the most universal. Here was an object manifestly in France and yet not in France, religiously redolent and yet stridently rational, absolutely material and yet completely abstract. In an age when "family, country, church" had become "family, country, science," **K** and **M** were perfect emblems of the Third Republic: buried in specificity, risen in universality. The symbolic resonance of the meter was lost on no one. Back in 1876, the Republic had even memorialized the new meter by striking a richly iconographic medal in honor of the standard, the scientists who were constructing it, and the glory of the original meter chosen during Germinal of Year III.[5] On the occasion of the sanctioning in 1889, French newspapers recalled with "patriotic satisfaction" how shortly after the "disaster of 1870" foreign scientists, even those who had previously impugned French precision, now acknowledged its triumph.[6]

Before the ink was dry on the Convention of the Meter, delegates were planning new standards that would be built on the model of **M**. Scientific-technical conventions not only garnered symbolic capital for the country or countries that spearheaded them, but they also engendered real benefits for trade exports and smoothed zones of national confrontation. Conventions were also responses to the sudden confrontation of industrial products at the international expositions, the commercial "chaos" to which Dumas had referred. But conventions also mediated the crossing of train lines and schedules, and blame rapidly fell on their absence when trains smashed into one another. For much of the early nineteenth century, regional (even national) systems of communication, production, and exchange had been free to grow in relative isolation. In the last third of the nineteenth century, systems collided at myriad boundaries in the colonies, markets, and fairs. It was this friction that the conventions were designed to ease. They were patches at the ragged fronts at which telegraphic, electric, and railroad networks met.

Figure 3.2 Burial of the Meter. *In the ceremonial 1889 "sanctioning" of the standard meter and kilogram at Breteuil (near Paris) the most painstakingly created physical objects were buried, so they could function externally as universal measures. Here* **M** *lies in its protective metal case on the upper shelf of its triply locked underground vault, while* **K** *presides over its six "witnesses," three standing to each side.* SOURCE: *LE BUREAU INTERNATIONAL DES POIDS ET MESURES: 1875–1975, P. 39.*

Governments drew up conventions to calm the frantic rustling of incompatible maps as navigators tried to plot routes at the boundary of colonial dominions. They introduced conventions to facilitate the movement of dynamos, gear trains, and steam engines. Regulating these confrontations required hard-fought instruments of accommodation, and their number multiplied: conventions of war; conventions of peace; conventions of electrical power; conventions of temperature, length, and weight. Conventions, as we will see, of time.

The years after Poincaré's January 1887 elevation into the Academy of Sciences came at the height of debate over these new stan-

dards. Academicians took an interest down to the detailed metal-
lurgy of the meter bars, and their fascination with the meter led to
further conventions, as when one of France's eminent astronomers
submitted a paper to the Academy in which the meter served as a
model for the decimalization of money. When, just after the sanctioning of the meter, one challenger wrote the Academy to dispute
the fidelity of the new bars to the old Archives standard, Berlin
astronomer Förster laid down the law: "The international committee of weights and measures [finds it] unacceptable to allow the
base of the metric system to depend on uncertain and incessant corrections, now that that base has been materially defined by the international prototype."[7] **M** now ruled alone.

Pushed by the French at every turn (in part out of principle, in
part as a countermeasure to the force of imperial Britain), the concept of *convention* widened, condensing into a single word a triple
resonance. *Convention* invoked the revolutionary Convention of
Year II that introduced the decimal system of space and time; *convention* designated the international treaty, *the* diplomatic instrument that the French, more than any other country, pushed to the
fore in the second half of the nineteenth century. More generally,
convention is a quantity or relation fixed by broad agreement. A convention, fixed by convention, in the tradition of the Convention.
When gloved hands lowered the polished standard meter **M** into
the vaults of Paris, the French, literally, held the keys to a universal
system of weights and measures. Diplomacy and science, nationalism and internationalism, specificity and universality converged in
the secular sanctity of that vault.

But if France could lock space and mass in the protected lower
basement of Breteuil, time proved more elusive. At the beginning
of the 1880s, one French review lamented that clocks were extraordinarily recalcitrant, each one's own "personality" repelling any
attempt to regularize it by making corrections based on temperature. Not that French astronomers and physicists had not tried. All

over Europe, neighborhoods, cities, regions, and countries were struggling to standardize and unify their clocks. In Paris and Vienna during the late 1870s, industrial steam plants injected subterranean pipes with compressed air, then modulated that pressure to set clocks pneumatically around the city. Customers could wander through pneumatic shops to select their preferred display of Victorian exactitude.

At first the fifteen-second delay caused by the time it took the pressure pulse to race under the streets of Paris seemed like nothing. Yet time sensitivity had sufficiently mounted by 1881 that even this tiny delay (causing the clocks at different points in the pipework to differ from one another and from the Observatory) became visi-

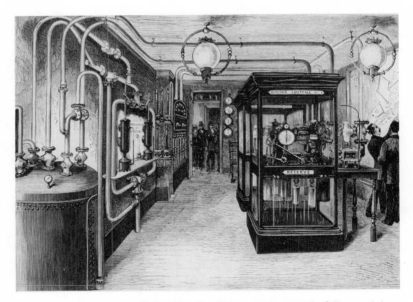

Figure 3.3 Pneumatic Unification of Time: The Control Room (circa 1880). From the control room at the Rue du Télégraphe in Paris, the pipelines pumped time under the city streets to synchronize clocks in every quarter of the metropolis. SOURCE: *COMPAGNIE GÉNÉRALE DES HORLOGES PNEUMATIQUES,* ARCHIVES DE LA VILLE DE PARIS, VONC 20.

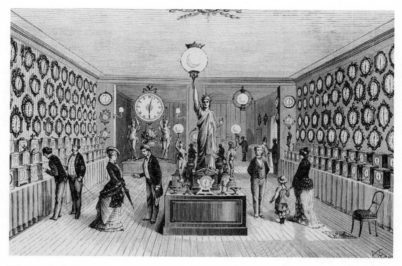

Figure 3.4 Pneumatic Unification of Time: The Display Room (circa 1880). Here customers—both commercial and private—could purchase clocks that would register the carefully timed bursts of air that they would receive through the pneumatic pipes of Paris. Source: Compagnie Générale des Horloges Pneumatiques, Archives de la Ville de Paris, VONC 20.

ble. Astronomers caught the problem, so did the engineers of bridges and roads. Soon the public did as well. At first the engineers tried to shrug off the discrepancy: "this small discordance, indisputable in theory, has little practical importance since we are only dealing with clocks that display minutes, and where the minute hands jump in steps and do not permit further divisions, even approximately between that division of time." The clock minders hastened to add that they would offset the Observatory's clock by the fifteen seconds the pulse took to reach the outermost reaches of the network. To be exact, they then mounted retarding counterweights on each pneumatic clock based on its distance from the center. In this way, they reassured their readers, "practically the whole of the discrepancy will be corrected."[8]

Two striking features of time coordination emerge from this little vignette. First, time awareness had become acute. Before the nineteenth century, clocks normally did not even have minute hands.[9] Now a fifteen-second discrepancy could drive engineers to modify public clocks. Second, the transmission time — even of a pressure wave traveling at the speed of sound — looked to professionals and the public like a problem demanding correction. But if the late-nineteenth-century public wanted their seconds adjusted, astronomers had long grown used to far greater precision. Urbain Le Verrier, director of the Paris Observatory and co-discoverer of the planet Neptune, had long since wanted electrical time unification. Synchronizing clocks by pneumatic means would have been absurdly inaccurate in the context of late-nineteenth-century astronomical work. In 1875, no doubt prompted by the Observatory's role in the unification of the system of weights and lengths, Le Verrier proposed standardizing and unifying Parisian time by electricity, as the astronomers had already unified the various rooms of their own observatory. Physicists Cornu and Fizeau, along with the Observatory's astronomers, all endorsed the idea. It was a perfect Polytechnique project. Le Verrier lost no time in pressing the Department of the Seine for support. Le Verrier and his astronomers insisted that their goal was to extend the interior order of the observatory to the whole of the city: "I propose to the City of Paris to give the public clocks a synchronized action and a precision superior to that which we have habitually satisfied ourselves. . . . If the City of Paris agrees . . . it will find here the opportunity to give a new and fertile boost to the art of clockmaking which has made famous the names of French artisans."[10]

Paris agreed, promptly establishing an illustrious commission to guide its clocks. Gustave Tresca would join the time standardization drive; it was he who was supervising the production of the standard meter sticks and weights that would grace the basement at Breteuil. Edmond Becquerel would be there, too, as a major French physicist (he was the father of Henri Becquerel of radioactivity fame). Renowned architect Eugène Viollet-le-Duc served on the commission, no doubt because of his famed restorations (coordinating grand church clocks presented huge architectural and structural issues). Charles Wolf, astronomer at the Paris Observatory, was a commissioner; he had invented much of the observatory's electric time-coordination system. The astronomers and their allies ran a clock-building competition and soon had a working trial system.

By the time the commission reported back to the City in January 1879, Le Verrier had died. But his plan lived. A dozen synchronized clocks would dot Paris, joined by telegraphic cable to the mother clock in the Observatory. Built precisely on the model of coordinated precision they had erected in their own Observatory, each of these secondary clocks had its mechanism set to run fifteen seconds fast each twenty-four hours. A controlling pulse from the Observatory drove an electromagnet in each public clock and that magnet slowed the pendulum, pulling the remote clock into synchrony with the mother clock. Each secondary clock radiated time electrically, resetting other public clocks in city halls, important squares, and churches. From now on, the report proclaimed, the public would have forty public clocks announcing the time correct to the nearest minute — indeed, to the nearest *second* just after it received the reset signal. Still, there were spatial and legal limits past which the Observatory time wires would not go:

> We have not included in the list of clocks to be regulated any belonging to the railroad. It is not that we have misunderstood the enormous interest that the public would have in knowing that these clocks are in agreement among themselves and with those of the City. But . . . it seemed to the Commission that it would be imprudent to engage the City . . . itself in such a complex service, where big interests are in play, and where its responsibility in case of accident arising from the regulation of the clocks could be

engaged in an unfortunate way. One can not doubt, however, that the [railroad] Companies, when they see at their doorstep all the clocks of the City regularly indicating the time, and all the same time, will spontaneously put themselves in accord with the time of the Observatory. That day, the unification of time in Paris will be the unification of time in all of France.[11]

Here was an admirable vision: the Observatory would stretch its walls until Le Verrier's system embraced the entirety of Paris. A clock at the country's center of precision would multiply itself until every jeweler, every citizen, would have astronomer's time within a stone's throw. By example, trains and finally all France would follow. Through this series of symbolic reflections—a temporal hall of mirrors—Le Verrier's astronomically set pendulum would set the time of every clock in the country.

The clocks never worked. Ice in the sewer systems promptly cut the wires at numerous points: the current ended up driving the clocks without the intervention of the mother clock. Soon public clocks all over Paris were hawking their own peculiar times. In embarrassment and anger, the commission attacked the chief engineer, precipitating a cascade of mutual recriminations over patents and the patent failure of the public clocks to register anything like the right time. Pleading that the commissioners use his latest inventions, the chief engineer lambasted clocks that were accurate only on receiving their reset signal: "In regarding the clockface at any moment, the observer must have the absolute *certainty* that the clock is correct to within a few seconds at the most, not *to within five minutes.*"[12]

During 1882 and 1883, reports streamed back to the authorities that the clocks of one arrondissement after another were not getting proper electrical guidance from the Observatory. By the spring of 1883 not a single public clock tied to the secondary regulators was receiving any current at all.[13] French authors conceded that their country had failed to dominate the time unification of cities. Adding insult to injury, it was London, home of the twelve-inch foot, that led the way toward standardizing time.[14]

After establishing the glorious, rational meter, it was galling to the French scientific establishment that synchronized time had slipped out of their hands. In 1889, the Observatory director pleaded with city authorities that this temporal chaos had to stop: "The Counsel of the Observatory, has been disturbed many times by the manner in which the distribution of time has functioned in Paris. The results obtained up till now are in effect far from being satisfactory, so much so that given the numerous protests, the director of the Observatory had in effect to request the erasure of any mention of 'observatory time'."[15] At the 1900 Universal Exposition, foreigners would see this sorry state of affairs. Couldn't the municipality and the Observatory build a system "more worthy of a city like Paris"? Under these circumstances, it became ever clearer that railroads were far from likely to mimic "spontaneously" the Observatory-City system, as Le Verrier had dreamed.

Times, Trains, and Telegraphs

It was not that the French railroaders did not want coordinated time. They, like the rest of Paris, were transfixed by the coming triumph of the Parisian standard meter as its 1889 sanctification drew near. The industrial *General Review of Railroads* opened its 1888 discussion of time by referring directly to the extraordinary success of metrical reform:

The metric system, one of the most glorious creations of French genius, has already conquered half of the world, and its complete triumph is no longer doubted by anyone. Its authors have added a new calendar, but they have not concerned themselves with fixing the beginning or middle of the day . . . questions which seem resolved by the advance of the sun. It required the rapidity of communication by rail and telegraph to seed the idea of choosing, more or less arbitrarily, the time of one locality to be imposed on others, and to create in this way normal or national hours. This has given birth to a confusion of a new kind but of the same type as that due to the multiplicity of ancient national weights and measures.[16]

In France, as in many other countries, each train system used the time of the main city served. Bit by bit, as lines from Paris wound deeper into the hinterland, they had chased away local times until, by 1888, Paris fixed the whole country's railroad time. Clock faces in the courtyards and departure lounges indicated the exact mean time of Paris, while platform clocks ran behind the outside clocks by three or sometimes five minutes to give the traveling public a margin of error. So as passengers waited in train stations outside of Paris — in Brest or Nice, for example — they experienced three times: their city's own local time, Paris time (in the waiting room), and an offset time in the track area. (Train time ran in advance of Brest by twenty-seven minutes and behind Nice by twenty.) The *Revue* analyzed other countries' time schemes, examining each one's solution to the time problem. Russia had unified time in January 1888. Sweden had set its clocks one hour later than Greenwich. Germany staggered under multiple *Land*-based times.

"Nowhere else has the question of time been posed in a more pressing manner than in the vast network of railroads of the United States and in the English possessions of North America." Grounded in the North American railroads' April 1883 decision to synchronize all their clocks by zones, the Americans and Canadians had chosen Greenwich as time zero, blocking out huge longitudinal swaths from "Intercolonial time" in the East to "Pacific time" in the West. "Let us add, before leaving America," the French railway journal concluded, "that the [American] charts and color maps offered to the public seem to us, by their clarity and beauty of their printing, noticeably superior to that which we usually see in our countries of ancient civilization." According to the *Revue*, when international scientific delegates gathered in Washington, D.C., in October 1884, it was the railroaders who had been able to remind them that "any change would be useless and inopportune." Now the stakes were clear: could the French—could the world—adopt a "generalized American system"? For the French railroad *Revue*, this was a question that should not be left exclusively in the hands of geographers, geodesists, and astronomers.[17] No doubt with Paris standoffish astronomers in mind, the author wrote: "It is only when the railroads and telegraphs have realized [time] reform that one can hope to see their example followed by other administrations and municipalities. And it is only then, as in North America, that the reform could be complete and could make felt its benefits."[18]

French railroaders, telegraphers, and astronomers looked with a mixture of admiration and anxiety at Britain and the United States when it came to time reform. America stood out for its industrial distribution of time, Britain for its world-dominating network of undersea cables. When Henri Poincaré joined the Bureau of Longitude in 1893, he entered a world quite different from the vast commercial and scientific enterprise administered by the British and the Americans. Clocks ran with stunning precision in the Observatory and appalling inaccuracy in the streets of Paris. The French, especially the Polytechnicians, rued this urban failure, but they were proud of their principled, mathematical, philosophical approach to standardization. They had brought the Enlightenment meter to a triumphant victory and begun extending the universal rationality it announced into the chaotic dominion of time.

On the other side of the Atlantic, North American time reform could boast no leader with the scientific stature of Le Verrier. It simply is not possible to reduce the American time coordination story to the work of an individual, an industry, or a scientist, despite many attempts. Instead, the movement toward synchronization was always critically opalescent, with dozens of town councils, railroad supervisors, telegraphers, scientific-technical societies, diplomats, scientists, and observatories all vying to coordinate clocks in different ways. That effort was so hybrid, so fluctuating in its allegiances and coordinated grids, that astronomers sold time like businessmen and railroaders spoke to the universal order of nature.

French savants found America's most impressive science not mathematical physics, mathematics, or pure astronomy, but rather the work of the ambitious Coast and Geodetic Survey. Teams of cartographers and surveyors were busy laying out the boundaries, rivers, mountains, and natural resources of the rapidly expanding country. Like all their fellow map makers, the Americans struggled with time, because time was inseparable from longitude.

Finding local time on the spot was a matter of watching the sky, then setting a clock by the moment when the sun passed its highest point. Or, more precisely, it meant determining the moment a certain star crossed an imaginary line running vertically up from the northern horizon. If the surveyors also knew what time it was back at a fixed reference point—Washington, D.C., for example—they could then simply reckon the time difference between local and Washington time. If the two times were the same, the surveyors were somewhere on the same longitude line as the Capitol. If the surveyors found their time to be three hours earlier than Washington, then they were an eighth of the way around the globe, to the west.

The map maker's problem was therefore always this same question of distant simultaneity: What time is it right now back in Washington or Paris or Greenwich? So explorers, surveyors, and navigators carried clocks (chronometers) set to the time of their port of departure. All the longitude finder had to do was to compare local time to the chronometer. But getting a precision clock to guard proper time in the unsteady motion of a ship's cabin or on a mule's back was never easy. Add the vagaries of temperature, moisture, and mechanical failings, and the provision of a stable, precise chronometer became one of the most difficult machine problems ever attacked. For John Harrison, the extraordinary eighteenth-century clockmaker, efforts to build an accurate seagoing longitude clock consumed the whole of his life.[19] Gifted though he was, Harrison did not end the search for movable time. The hunt for reliable, transportable clocks continued throughout the nineteenth and twentieth centuries. Astronomers fought long and hard to devise a precise way to use the moon's movement against the fixed stars as a giant clock readable from anywhere. But the moon's position was hard to fix mathematically, and in the field or on a ship it was difficult to measure just where the moon was, except for those rare moments when it actually passed in front of a star or planet.

The single measurement American surveyors most wanted was the longitude difference between the New World and the Old. But the map makers simply could not come to a consensus. One desperate series of attempts—only one among many—began in August 1849, with seven transatlantic voyages in each direction, each bearing twelve accurate chronometers. The hope was that their time cargo would finally show the true difference in time, and therefore longitude, across the Atlantic. In 1851 they stashed on board thirty-seven chronometers, taking advantage of five sailings from Liverpool and two from Cambridge, Massachusetts. After hauling ninety-three chronometers across the seas, the astronomers optimistically claimed a shore-to-shore time difference to within one-twentieth of a second.[20]

Such vaunted precision soon rang hollow. Despite the presence of ever-more-vigilant clock tenders on the ships to protect the ticking freight, the measured time difference from the United States to England was, impossibly, different from that from England to the United States. Something on the high seas was confounding the clocks. Astronomers suspected that temperature was probably the culprit, with the lower temperatures far offshore slowing the clockworks. This meant that if a lethargic clock set sail at 1:00 P.M. from Cambridge, Massachusetts, it would arrive in Europe showing that Cambridge time was earlier than it really was and induce British map makers, relying on the clocks arriving in England, to put Cambridge, Massachusetts, to the west of its actual location. Conversely, a slow-running seagoing clock set initially in Liverpool would suggest to the Americans that Liverpool time was earlier than it was, so the New World map makers would draw their maps with Liverpool to the west and therefore *closer* to the shores of North America. Testing, calculating, and interpolating did little to help. Crowding the ship with more supervisors, better temperature compensators, and superior clocks just spewed out additional conflicting data. Unable to measure the number that mattered most, the longitude difference between North America and Europe, map makers despaired.

For hundreds of years, cartographers could only dream of being able to send a signal of simultaneity to fix longitude. The telegraph cracked the problem. Over vast distances, an electric current would race a signal through the wires so fast that the reception and transmission seemed practically instantaneous. During the summer of 1848, observatory astronomers from the Harvard Observatory and the Coast Survey tested this new function for the telegraph. One person would tap on the key and the other would listen for a beat at the distant end. Each distant tap would leave a mark on a paper that wound through the receiver's printing device. One evening, one of the mappers, Sears Walker, wondered aloud if they couldn't directly observe and transmit their observation of a star passing through the north. Bond replied: Why not make the escapement of a clock act like a telegraph key so the ticks would be heard anywhere and everywhere along the telegraph line? Then, Why not let that clock-driven signal mark on a smoothly turning cylinder located far from the clock?[21] By comparing the position of marks made by a locally set clock with the position of marks made by signals sent at known times from a distant clock, surveyors could accurately compare distant and local time.

distant 12:00:00→| distant 12:00:01→ | distant 12:00:02→|
local 12:00:00→|

For instance, here the local noon occurs about a half-second after the distant noon. Instead of trying to register time by stopping a clock when a star crossed a spider-thread reticle in their telescope, the astronomers could simply measure the distance between lines on their paper. From that simple measurement, the surveyors had longitude.

By the end of 1851, telegraph cables stretched from Cambridge, Massachusetts, to Bangor, Maine; from there the time signal jumped, in a second transmission, from Bangor all the way to Halifax, Nova Scotia. As American scientists began advertizing their electric time transmitters, they found a ready audience in Europe. Bond noted: "It is a gratifying circumstance that this invention is known and spoken of in England only as the 'American method,' and the Astronomer Royal has laid the wires at Greenwich preparatory to introducing it there."[22]

It was not just the astronomers and map makers who cared about the rapid dispersal of simultaneity. Trains had schedules to maintain, and by 1848–49 railroads began forming voluntary associations to fix, by convention, the time on which they ran. For much of New England that meant that all trains on or after 5 November 1849 were to adopt the "true time at Boston as given by William Bond & Son, No. 26 Congress Street."[23] Any railroad not already in this common time system soon was motivated to be so. On 12 August 1853, two trains of the Providence and Worcester line slammed into

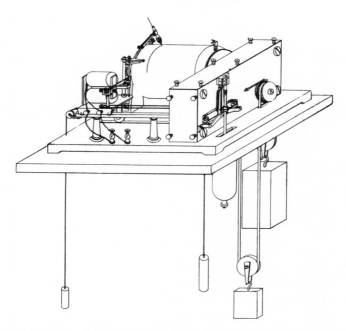

Figure 3.5 The American Method. *By recording the arrival of telegraph signals on a precisely rotating drum, the transmission of time could be made vastly more accurate than by earlier, acoustic means. For longer stretches (under the Atlantic, for example) the simultaneity men used the more-sensitive method invented by Lord Kelvin: the incoming electric time signal caused a mirror-mounted magnet to twist ever so slightly, which caused a reflected light beam to shift on a sheet of paper.* SOURCE: GREEN, REPORT ON TELEGRAPHIC DETERMINATION (1877), OPPOSITE P. 23.

each other at a blind curve. Fourteen people died, and newspapers blamed the tragedy on a conductor with an itchy finger on the throttle and a slow watch at his side. With another bad-watch disaster just a few days earlier, train lines found themselves under immense pressure to coordinate their clocks. Telegraphically transmitted time became a standard railroad technology.[24]

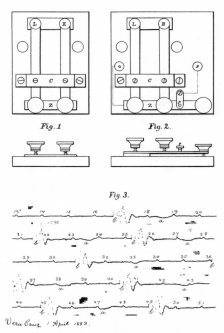

Figure 3.6 Traces of Time. *Telegraph keys and the traces of distantly sent time signals recorded by the "American Method" (1883).* SOURCE: CH. HENRY DAVIS ET AL., *TELEGRAPHIC LONGITUDE IN MEXICO AND CENTRAL AMERICA* (1885), PLATE 1.

Partly pushing the observatories, partly pushed by them, train supervisors, telegraph operators, and watchmakers accelerated the electrical coordination of clocks both in England and in the United States. By 1852, directed by the Astronomer Royal, British clocks were sending electrical signals over telegraph lines both to public clocks and to railways.[25] Soon the Americans were, too. Summing up the status of their time effort, the director of the Harvard College Observatory boasted in late 1853 that the "beats of our clock can, in effect, be instantly made audible at any telegraph station within several hundred miles of this Observatory."[26] During the 1860s and 1870s, coordinated time reached deeper into the cities and train system. Hailed in the press, visible in the streets, studied in observatories and laboratories, the synchronized clock was anything but rarified science. Its capillary extension into train stations, neighborhoods, and churches meant that synchronized time intervened in peoples' lives the way electric power, sewage, or gas did: as a circulating fluid of modern urban life. Unlike other public services, time synchronization depended directly on scientists. By the end of the 1870s, the Harvard College Observatory was but one site sending time, though for a few years its service was one of the largest. Idiosyncratic developments in Pittsburgh, Cincinnati, Greenwich, Paris, or Berlin set each apart.[27]

Marketing Time

Shortly after those first experiments on electrical time, Harvard Observatory hired a telegraph line to distribute to Boston the time its astronomers had determined by the measurement of the stars. By 1871, the observatory director was charging for the service, hoping to install a prominent clock "so that the public can see and learn to appreciate the method of communicating time."[28] Returns were good: in 1875, the service netted $2,400, a yield sufficiently large that the observatory hired a proprietor for its time business.[29] Astronomer Leonard Waldo came to Harvard in February 1877 from Yale, where he had run a similar service. By then the Cantabrigians had invested over $8,000 in instruments, clocks, and telegraph lines. Now they needed customers. Using their telegraph to trigger its noon start, Waldo planned to drop a large copper time ball down a mast on top of one of Boston's higher buildings. He hoped that this public display, visible both to landlubbers and navigators, would dramatically boost the observatory's recognition. So would a major recruitment of railways. Jewelers and clockmakers too were needed as time clients, not to speak of private individuals who clamored (or ought to, according to Waldo) for precision time.

Waldo hoped a widening clientele would persuade big manufacturers that time really was money, or at least was worth buying.
Wires multiplied, snaking through the observatory that aspired to
be the master clock for all New England.[30]

Hard-driven Waldo pounded out his message wherever he could.
Encouraged by Western Union, he printed a pamphlet that adver-
tized the time service to a hundred New England cities and towns.
His time, Harvard Observatory's time, would be based on an impressive Frodsham clock, corrected once each day at 10:00 A.M. to
within a fraction of a second. From this master clock, time would
course through the region on two circuits in addition to the main
one linking the observatory, the Boston Fire Department, and the
city of Boston. In the unlikely event that the Observatory clock
failed, Waldo promised his time-hungry customers that they would
receive a back-up signal from clockmakers William Bond & Sons.
All day long, at two-second intervals, pulses shot out from the
Observatory, skipping, for identification purposes, every minute's
fifty-eighth second, while every fifth minute the service would omit
the pulses that would normally go out on the thirty-fourth to sixti-
eth seconds.[31]

Citizens like the proprietor of the Rhode Island Card Board
Company wanted to know the time:

> Will you have the goodness to inform me whether the time fur-
> nished at the W[estern] U[nion] Telegraph office Boston from
> your observatory is actual Cambridge time, or Boston time. — That
> is, whether an allowance is made of the difference in time between
> the two places before transmitting the signals — [.] Does the little
> hand that beats seconds on the dial in Boston correspond precisely
> with the beating of the standard clock at Cambridge? Are correct
> signals furnished to Providence by telegraph direct from your
> observatory? — A number of Citizens of Providence and vicinity
> having fine watches are desirous of having this information.[32]

Did cardboard manufacturing demand split-second timing? Rather, the cardboard mogul and others like him wanted exact time because their "fine watches" demanded it; they had come to view the exact coordination of their timepieces as a value not captured by the purely pragmatic. Fostering such a modern temporal enthusiasm was Waldo's great hope. Not only his own circulars, talks, and correspondence but surely also the ever-increasing role of railroads had begun to create a sensibility that could only work to the advantage of Harvard's time service. Five years of selling time had done much to foster this awareness. Waldo put it this way:

> In . . . time the community generally had been unconsciously educated to the desirability of a uniform standard of time. Accurate time-pieces, in many places, were daily compared with the Observatory signals, and their rates so determined were considered authoritative. So critical, indeed, had the subscribers become, that errors of a fraction of a second were detected and commented upon.

Did the running of trains and the calibration of fire bells require an accuracy that would better an error of four-tenths with one of two-tenths of a second? Of course not. Yet Waldo both pushed the public and was pushed by them to ever greater precision. In his own way, he participated in the creation of a modernist time sensibility, one with bounds that far exceeded the practicalities of precision.[33]

Back at his time factory, Waldo struggled to protect his regulator clocks from variation in climate, to systematize telegraphic connections, and to engage a dedicated observer to determine clock error each day. To guard the university clocks from variations of temperature, the astronomers sealed a basement room against the elements. This "clock room" stood in the west wing of the observatory and was completed 2 March 1877. Some 10' x 4'2" wide, 9'10" high, its double walls protected the all-important clocks. Only a safe door penetrated these thick barriers. When the door closed snugly it sealed against felt packing. Inside, each of the three treasured clocks stood on marble slabs on brick piers, their faces illuminated by tin reflectors so they could be viewed from behind small, thick glass windows (immediate human presence posed a threat to accuracy). In observatories around the world, from Berlin to Liverpool, from Moscow to Paris, similarly obsessed astronomers shaved errors from their mother clocks.[34]

Split-second timing was one thing; marketing it was another. At the limit of the Boston time-selling region stood Hartford, a city that had vacillated between its own local time and that of New York City. When a Hartford worthy, Charles Teske, wrote to Waldo in July 1878, indicating his interest in converting his town to Harvard Observatory time, Waldo took notice. Teske reported that he had corralled the Hartford Fire Department and its committee, fire departments typically being the agency whose noon bell signaled time to the public. Hartford's mayor liked the idea of purchasing Cambridge time to set hammer against bell at noon and midnight. But rounding up diverse interests and divergent times was less simple, as Teske lamented to Waldo: "It is like awakening the dead out of their sleep, to get people interested in this matter, for we have here in Hartford all sorts of time and everybody claims that his time is correct." Teske reckoned it would cost him a good deal of effort, and a reasonable price from the observatory, to nail down the board of aldermen and the common council. But more than just money, autonomy was at stake. "Will you give us Cambridge time? Can you give us Hartford Time? What is the exact difference between Cambridge & Hartford?" Several months later he was still hammering away at his various opponents.[35]

Scratching out his report in November 1878, Waldo left a blank where Hartford would be, he knew that city stood on the regional border between two time dominions carved by train tracks. "It is to the Rail Road that we must look for the most certain support of our scheme for a general distribution of the observatory time signals. . . . it is the rule that the Rail Road regulates the time of towns along its line." Train time followed the big city centers, and the New York to Hartford line put the latter squarely in New York's time radius. But the domain from New London and Providence up to Springfield fell in Cambridge's time zone. "For some years to come therefore I think Hartford will be the point at which we must cease to give Boston time." Disputed lands might lie at the margins, but the regional conquest of simultaneity was never in question: "local time throughout New England should be discouraged."[36]

Cambridge lost Hartford. Even Teske's desperate recruitment of the president of Trinity College could not move the city council. Eventually the Hartford Fire Department threw up its hands and bought a marine chronometer to ring its noon alarm. Dejected, Teske scribbled to the observatory that "of course" this was "considered good enough by those who are ignorant of the minuteness of your Time Signal." Having failed to move his city, Teske offered to buy Harvard time for his store (one assumes he sold timepieces); he was determined to have an "absolutely correct Time Signal." Two years later, Connecticut officially set its time to the meridian of the city of New York using electrical signals from Yale Observatory. From the college, time would siphon through the New Haven junction up the railroad tracks of various railroad companies. These train lines were in turn obliged by the law to shoot electric time down all their roads, to mount coordinated clocks at their stations, and to forward time to all intersecting railroads.

Railroad magnates were not amused. They had their own times and resented any state intrusion. The general manager of New York & New England Railroad Co. grumbled, "The standard time of this road is Boston time . . . but under a ridiculous law in the state of Connecticut we are obliged to use N.Y. time whenever we cross the state line. . . . It is a nuisance and great inconvenience and no use to anybody as I can see."[37]

Throughout: sudden changes of scale. One day the Harvard Observatory would solve a navigator's problem. Another day, its staff worried about the frozen time ball they released down a high mast by remote control. On yet another, they drafted plans to wire hundreds of cities across the whole of New England. All around the United States, such regions of time coordination grew, as did friction at their boundaries. A jeweler's store here, a train line there; occasionally a meridian section, a region, or a state. By the end of 1877, Dearborn Observatory (south of downtown Chicago) controlled the clocks of half a dozen jewelers, four main railway companies passing through Chicago, and the Chicago Board of Trade. If Harvard's Waldo was fanatically striving to shave hundredth-of-a-second errors off his regulator, Dearborn cheerfully indicated that "no great pains are taken to keep *very* accurate time. The standard signal clock is generally kept within half a second of the exact time. . . . As long as the time furnished is within a second or so of being exact we are satisfied, and we consider that close enough for all practical purposes."

Dearborn was right. Train operators and passengers surely did not need clocks more accurate than human reaction time. But differences in time culture reflected more than that. First, Harvard Observatory's time service had its origins in longitude determination. There a wide consensus existed, not just in the United States, that map making ought to be as accurate as possible. Surveyors wanted accuracy to tenths if not hundredths or even thousandths of a second. Second, the cultivated precision mania of the Harvard Observatory had come to resonate not only among astronomers but also among high-end watchmakers and their gentlemanly New England customers. The culture of precision ticked sooner and louder in Boston than in Chicago.[38]

But the greatest contrast lay not among the American observatories but between the United States (or Britain) and France. It is simply unimaginable to picture an American or British astronomer in 1879 refusing to pull a time-bearing telegraph wire through the door of a railroad station, as unimaginable, certainly, as it is to picture an American astronomer defending the unification of time because it would complete the unfinished business of the Age of Reason.

Measuring Society

In the fluctuations of scale that marked every stage of the time campaign, one powerful surge toward the global came in the 1870s through the American Metrological Society, founded by Frederick A. P. Barnard, president of Columbia University. Calling on an illustrious assortment of scientists, he urged an internationalism that grounded its cosmopolitanism on commercial exchange: "The diversity of the conventional methods employed by different peoples for determining the quantities and values of material things have been in all times a source of infinite embarrassment to the operations of commerce, and a serious obstacle in the way of intelligent inter-communication between nations." Metrological reform on the Continent struck Barnard as a sign of hope, one that had yet to cross the shores of the English-speaking lands. Remedying that failure was the goal of the society. On Tuesday, 30 December 1873, the group approved a constitution that set out to "bring [weights, measures, and moneys] into relations of simple commensurability with each other." Metrologists would attack the zero of longitude, the units of force, pressure, and temperature, as well as the quantification of electricity, bringing their results to the attention of Congress, the states, boards of educations, and universities. It was, in short, a lobby for focusing French rationalism through a commercial lens and printing it on American civil society, from the schoolroom to the railroad yard.[39]

Cleveland Abbe, meteorologist and astronomer with the United States Signal Corps, found his way to the cause through frustration with patchwork time. In 1874, when Abbe had enlisted amateur observers to study aurora, his troops couldn't agree on a common time base, confounding Abbe when he tried to combine the various field sightings. Appealing to the metrological society for help, with bureaucratic justice he was promptly made chair of the committee on coordinated time. With Barnard and the Canadian booster Sandford Fleming, Abbe became one of the most outspoken lobbyists for time unification, producing material used before Congress in 1882 and at international congresses in Venice (1881) and Rome (1883).[40]

Abbe's metrological society time team argued in 1879 that truly local times, in an astronomical sense, had long vanished in favor of a mishmash of railroad times. Consequently, the society urged "public institutions, jewelers, town and city officials . . . to regulate the public clocks, bells, and other time signals, by the standard adopted by the principal railroads in or near their respective localities." One suggestion was to strip the seventy-five various meridians down to three zones, corresponding to four, five, and six hours west of Greenwich. Such a partial unification, they judged, would be good. But *one* single time standard for the nation would be far better. Set six hours west of Greenwich, this fully *national* hour would be dubbed "Railroad and Telegraph Time" for "these corporations exert such an influence on our every day life that we are persuaded that if they once take that step towards unification about which they have been talking for many years, then every one in the whole community will follow."[41] Precisely at the moment when Le Verrier was urging French railroads to adopt astronomers' time "by example," Americans like Barnard and Waldo wanted civil time modeled on train time.

Stridently modern, the metrological society resolved to lobby both state and national governments for changes that would extend deep into the fabric of ordinary life. Observatories, allied with railroads and telegraph systems, could deliver time anywhere on the continent. The metrologists insisted that railroad and telegraph time be displayed by a coordinated clock in every public building, every national and state capital, every hospital, prison, custom house, mint, life-saving station, light house, navy yard, arsenal, post office, and postal car. "On and after July 4th, 1880," they wanted *one* standard legal time for all purposes throughout the United States.[42] To win this battle, the metrological society knew it would have to recruit railroad and telegraph officials as well as astronomical allies, including Waldo and his counterparts at other observatories. As important a recruiting target as any was the thirty-three-year-old secretary of the General Time Convention of Railroad Officials and editor of *Travelers' Official Railway Guide for the United States and Canada*, William F. Allen.

Allen was the time shuffler of all time shufflers, the bureaucrat behind the endless lists of trains and schedules. It was Allen who scheduled the local schedulers; Allen, the railroad man who, more than any other, stood responsible for organizing the myriad iron roads into a guide that allowed the traveler to make connections between lines. And he, at least in June 1879, equivocated before he came to back a plan for unification.[43] On the one side, he was being pressed by the scientists' continental scale of action: they wanted the whole country under the rule of a single time convention. On the other, he and the railroads had to accommodate ordinary people who were used to operating on local time. So he compromised, settling for the status quo that nailed time along the tracks to the big cities they served. In June 1879, Abbe dismissed Allen's antique attachment to calling it twelve o'clock when the sun happened to be near the meridian. "The advantages of uniform time far outweigh the first week's awkwardness of the proposed change. Abolish old times is the watchword."[44]

Time reformers popped up everywhere. One outsider, a teacher named Charles Dowd, unsuccessfully petitioned Allen to adopt his zone system.[45] If Dowd was an outsider to railroading, looking in,

Sandford Fleming was the consummate insider, a railroader and promoter, looking out. The Canadian engineer took to massive projects with an eye scaled to empire. Having cut railway lines through the maritime provinces, put forward plans for the Pacific railway, co-designed Toronto's Palace of Industry, set the beaver on Canada's first stamp, and later lobbied for the transpacific cable, Fleming proposed a system that would cover the world. He had no time for town council pragmatism, but instead entered with a progressivist, imperial swagger, and a tincture of injured colonial pride. In 1876, he began writing articles hailing the new train and telegraph technologies, while deriding as antique the time systems they shattered: "We still cling . . . to the system of Chronometry inherited from a remote antiquity, notwithstanding difficulties and inconveniences which are constantly met in every part of the world." Fleming imagined today's traveler heading by ship from London to India. Hardly had this modern expedition left the shores of England, and the voyager's time was wrong. Paris time displaced Greenwich time, only to fall to the times of Rome, Brindisi, Alexandria and then to new times assigned daily until the ship puts into port in India. Maddeningly (in Fleming's view), Bombay time split between local and railway, with the railway time pinned to Madras. Such regressive confusion, Fleming argued, had to be simplified.[46]

Fleming wanted a single, universal time convention for the earth as a whole, with each of the twenty-four zones labeled by assigning to the 0-degree longitude line the name prime meridian; the letter A would stand for the 15-degree line, B the 30-degree line, and so on through X (345-degree line). "Universal," "cosmopolitan," or "terrestrial" time was then determined by an imaginary clock fixed in the center of the earth, with its hour hand pointed permanently at the sun. When the earth rotated so that meridian line C crossed the hour hand's line of sight to the sun, it would be C o'clock *everywhere*. When D crossed that imaginary line, it was D o'clock everywhere. *Every* clock in the world would display the same time simultaneously. If Big Ben said it was C:30:27 (meaning 30 minutes and 27 seconds past C o'clock), then so would a clock at Times Square or downtown Tokyo, the "electric telegraph affording the means of securing perfect synchronism all over the earth." Here was a scale of electric time unification that dwarfed even continental America. To accommodate local sensibilities, Fleming hatched watchmaking schemes. One allowed the user to rotate the whole watch face so the wearer's local noon (say, for example, F or Q, depending on location) stood at the top. Another design provided side-by-side clock faces, one for local, the other for terrestrial time.[47]

To Fleming, such time coordination was vital for the larger countries like Canada, the United States, and Brazil; it was helpful for the European nations of France, Germany, and Austria, and it would obviously benefit Russia, with its 180-degree longitudinal expanse. But Fleming's Canadian vision still had its focal point on London: "It is of still greater importance to the Colonial Empire of Great Britain with its settlements and stations in nearly every meridian around the entire globe, and with vast territories to be occupied by civilized inhabitants, in both hemispheres." Railroads and telegraph lines drove this unification. Some 400,000 miles of telegraph lines now ran over seabed and land; 95,000 miles of train track sprawled through Europe and Asia. Railway men like Fleming expected that the world would soon boast a million miles of rail alongside even greater lengths of wire.

Lines of telegraph and steam communications are girdling the earth, and all countries are being drawn into one neighbourhood—but when men of all races, in all lands are thus brought face to face, what will they find? They will find a great many

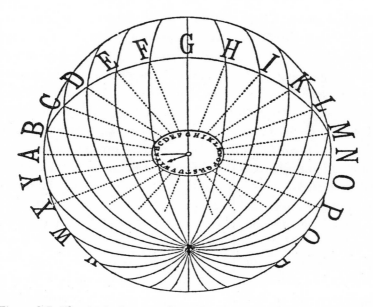

Figure 3.7 Fleming's Cosmopolitan Time. *Having begun as an engineer carving railroad routes across Canada, Sandford Fleming fashioned himself as a great defender of a system of electrically based time that would link the whole world to a single "cosmopolitan" hour. In this 1879 diagram, he showed the world itself as a giant clock for which twenty-four letters would replace the traditional reckoning by 1, 2, or 3 o'clock. "C o'clock," he hoped, would become the universal time when an imaginary line from the center of the earth to the sun passed through the longitude line conventionally labeled "C."*
SOURCE: FLEMING, "TIME RECKONING" (1879), P. 27.

> nations measuring the day by two sets of subdivisions, as if they had recently emerged from barbarism and had not yet learned to count higher than twelve. They will find the hands of the various clocks in use pointing in all conceivable directions.

This chaotic condition, Fleming insisted, had to end.[48]

Fleming's first article "Terrestrial Time" failed to reach the

powerful audience he imagined. In 1879 he struck again, recy-
cling his first attempt but now in even broader terms, pushing
more explicitly than before the need for a universal prime meridian (at Greenwich). The dustbin of history was full of failed con-
tenders for this line, he noted. Cape Verde (about 5 degrees west
of Senegal) had been one such putative zero meridian. Gerard
Mercator had fixed on the Island del Corvo in the Azores because
the magnetic needle there pointed due north. The Spanish had
chosen Cadiz, the Russians Pulkovo (outside St. Petersburg), the
Italians Naples, the British Cape Lizard (Cornwall), while Brazil-
ians referred their world back to Rio. If one put the prime meridian at the greatest human construction, then it should cross the
Great Pyramid—or so claimed the mystical Scottish Royal
Astronomer Piazzi Smyth.

But when Fleming had finished with his ecumenical homage
to prime meridia of years past, he returned to Greenwich, from
which nearly three-quarters of the world's shipping set its course.
Happily, he reported, the British *antimeridian* (the line 180
degrees removed from Greenwich) ran through the Bering Strait,
crossing a tiny bit of Kamchatka but otherwise spanning water
from pole to pole. Setting the prime meridian in the Bering Strait
would leave the world's (Greenwich-based) longitude lines
untouched and would necessitate only a minor adjustment in the
labeling of those lines. Excoriating the French for their national-
istic preference for Paris, Fleming laid the "universal" line
through the observatory central to the British empire. Shipping
and imperial power trumped history, mysticism, celestial mechanics, and other nations' nationalism.[49] Fleming's proposal fell like
music on the ears of American time reformers. Abbe informed
him in March 1880 that he was working to enlist telegraph and
railroad companies; Barnard proudly displayed Fleming's
cosmopolitan-time watches.[50]

Not everyone was sanguine about this new universalism. Barnard alerted Fleming that Piazzi Smyth was advancing contrary ideas. In 1864, Smyth had decided that the Great Pyramid embodied the wisdom of sacred Hebrew metrology, ancient knowledge that held contemporary relevance: "seeing how our nation is agitated just now by questions of a change in its hereditary Metrology, and is urged by a powerful political party to take a radically subversive, instead of a correcting, improving, and reforming, step in that direction." Where French (and some British) metric reformers saw progress, rationality, and universalism, Smyth saw gestures "fatal to national interests and connections." Where British units struck metric advocates as irrational, Smyth loved the inch and cubit as the last links via the Pyramids to ancient divine light.[51]

Barnard blasted back. Given the balance of political and cultural force, the North American reformers judged themselves able, more or less, to ignore Smyth's opposition. But British Astronomer Royal George Airy could not be so breezily toppled. So when Airy weighed in against Barnard in July 1881, it must have stung (Barnard immediately forwarded a copy of the letter to Fleming). Airy addressed Barnard politely but would not deign to mention Fleming by name. "It seems to me," Airy cautioned Barnard, "that you must begin with considering, on grounds of convenience and inconvenience, what *the mass of people want*; and must think of means of supplying that want." Whatever "the Canadian writer" might think of the traveler's travails, the issue of resetting one's watch on long-line railroads was not a problem, claimed Airy. Let the San Francisco–bound train from New York set its clocks by New York time; on the return trip passengers and railroad men could simply abide by San Francisco time. No, the practical problem was not on long American trips, but rather in the "limitohoptious" districts straddling lines of time change. "As to the *Cosmopolitan Time* what does a man living in Ireland or Turkey care about Cosmopolitan Time: It is wanted by sailors, whose profession carries them through great ranges of longitude. . . . And there its utility ends." Incredibly to Barnard and Fleming, Airy judged Greenwich itself a poor choice for a time center. Too far to the east to be fairly located within England, the observatory's sole claim to legitimacy came in being authoritative. For Airy, time should be unified only where unification made sense, as in the island of England. But trying to establish a universal time was a futile battle for the inconvenient goal of "hard and fast" time lines. "I do not imagine that it will ever be received."[52]

Writing to Fleming, Barnard tried to soften the blow by referring to Airy's excessively "opinionated" views and occasional past blunders, including his very public and wrongheaded intervention in the contentious dispute between French and British claims to the discovery of Neptune. Barnard reported to Fleming that he had lobbied Colonel Clarke of the Ordnance Survey in Britain with rather more success; Clarke would support the cause of time unification. But it was clearly a delicate moment. All Barnard's political skills were tested as he tried to get negotiators of the international Law of Nations to endorse time reform. Barnard lost the chairmanship of the time committee to the vastly more famous William Thomson (Lord Kelvin), who would, Barnard confided, need to be "educated." Worse, no one had bothered to notify Thomson that he was on the Law of Nations time committee, much less its chair. But while Barnard was trying to inject enthusiasm into Thomson, he had to dampen Fleming's boosterism, pleading with his northern ally in letter after letter to reign in his scatter-shot polemics: "I esteem it bad policy to evince uncertainty on the eve of battle, or to change point in face of the enemy." Time languished in committee.[53]

The apostles of unified time faced difficulty back in North America, too. While Barnard and Fleming stumped for a unified international time with Greenwich as its center, the United States Naval Observatory aimed for one fixed to the nation, not the world. Naval astronomers scoffed at the idea of ordinary people gaining anything whatsoever from a worldwide time, and opposed the locality of time zones. On the contrary: they wanted a European-style scientific time, grounded in a national observatory (theirs), uniform for the entire country. Rear Admiral John Rodgers, superintendent of the observatory, girded for battle in June 1881, looking skyward for support: "The Sun is the national clock used by many, and its position regulates the hour of rising, eating, working, and of going to rest. No other clock can supersede it, as it is the one ordained by Nature to regulate man's life." True, he conceded, railroads needed their own time, and the federal government could mandate the printing of Washington time in the schedules. But "[t]he people who do not care for scientific time are a thousand for one of those who do, and besides, I see no overpowering reason why we, with fifty millions of people should take the scientific time of a nation [England] with only thirty millions. If numbers and growth prevail, they should accept ours. I think that the feeling of nationality is too strong with the masses for philosophers to talk it away." Scientists, Rodgers concluded, "sometimes overestimate their functions."[54]

In the increasingly time-wired world of the early 1880s, time reformers campaigned for time unification on conflicting scales. Barnard, Fleming, and their allies pushed for a globe-covering "Terrestrial" time; the great national observatories of France, Britain, and the United States each advocated its own national time. Railroads and cities were the wild cards, much depended on the conventions they chose. Would cities conform to train time as they did in Britain and much of America? And if so, by what geography of simultaneity? Or would trains retain their own time, as in France? Convenience and convention were the watchwords, but whose master clock would set which lesser clocks? In late 1882, Allen endorsed a compromise proposal: set one-hour time zones for the whole North American railroad system. The metrological society welcomed Allen's efforts, enthusiastically reporting that Congress resolved to instruct the president to call for an international convention to which the U.S. would send three delegates. New York was contemplating switching its time ball to Greenwich time, the U.S. Post Office adopted a common time, and Barnard was lobbying President Chester A. Arthur while stumping for time reform with Secretary of State Frederick T. Frelinghuysen. Both the American Society of Civil Engineers and the American Association for the Advancement of Science backed the effort.[55] On the railroad tracks, the time unification scheme rolled even more swiftly. Some southerners briefly rebelled with a system tailored to their needs; but their move was crushed.[56] When Allen and the railroad presidents, managers, passenger agents, and superintendents met for their General Time Convention on 11 April 1883 in St. Louis, Missouri, railroad sentiment seemed to have turned toward the change to an hour system divided along the 75th, 90th, and 105th meridian lines.

Allen weighed in hard: ". . . the 'Hard Scrabble' system now in use, with its fifty different standards intersecting and interlacing each other, is an abomination and a nuisance which cannot be too soon remedied." To displace such chaos Allen presented his *pièce de résistance*: a map with three longitudinal time zones, each tinted in its own color, and its enemy, a multicolored quiltwork of the current division of time. Enthusiasts could buy either map, though the second one (depicting current chaos) couldn't be mass printed (too complicated) and cost twice as much. "A single glance at this map is sufficient, I think, to convince every one of the absurdity of the present status of this matter." As Allen made clear, the time zone system could have its zero line anywhere. True, the temptation would be great to pin the zero of longitude in Washington, D.C. But such parochialism was, he contended, unacceptable:

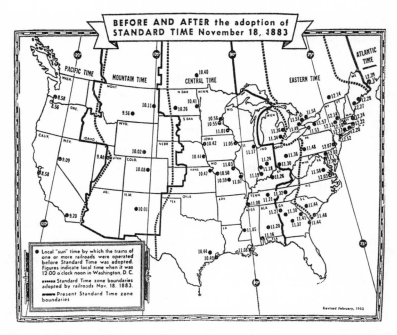

BEFORE AND AFTER the adoption of STANDARD TIME November 18, 1883

Local "sun" time by which the trains of one or more railroads were operated before Standard Time was adopted. Figures indicate local time when it was 12:00 o'clock noon in Washington, D. C.

Standard Time zone boundaries adopted by railroads Nov. 18, 1883.

Present Standard Time zone boundaries

Revised February, 1952

Figure 3.8 Trains, Times, and Zones. Time unification—an issue for reformers, astronomers, and standardizers—gained power when railway schedulers took up the cause of zoned time to avoid the proliferation of local times. This railway map plots many of the different local times before the zone reform of November 1883, as well as the dividing lines that the railways adopted after that date. SOURCE: CARLTON J. CORLISS, THE DAY OF TWO NOONS (1952), P. 7.

We are, all of us, more or less imbued with a feeling of local pride, and if the meridian time of the "Hub of the Universe" is the standard by which the trains on our particular road are run, we feel like holding to it. But, my friends, what right have you to claim that particular meridian as belonging to your city. The villages of Gum Tree and of Hard Scrabble are on the same meridian and have as much right to give it a name as your beautiful city has. . . . For all ordinary business transactions one standard is as good as another, so long as all agree to use it.[57]

Time was a convention, an agreement like any other that would, depending on the accord, unify cities, lines, zones, countries, or the world. Inscribing that arbitrariness into the collective language was as great a transformation as the acquisition of a regularized time awareness.

Both astronomers and railroaders viewed the new technologies of transport and communication as disciplining time more effectively than any school. As Allen put it, "Railroad trains are the great educators and monitors of the people in teaching and maintaining exact time." Train lines had altered the experience of time across Europe and North America; more than that, for an ever-growing portion of the population, railroad schedules had come to define time, to instantiate synchronicity. Indeed, without the quintessentially modern trains and telegraphs, the temporal structure of the world would, for most people, drift from its moorings. "I venture to assert," Allen added, "that if this city were cut off from railroad and telegraph communication for an entire month, and on the first night all clocks and watches were simultaneously and surreptitiously set half an hour faster or slower, not one person out of a thousand would . . . discover for himself that any change had been made."[58]

Winding up the convention, Allen slipped into a revealing parable that was at once tongue-in-cheek and reflective of the railroaders' position, since in setting conventions of time they were determining the electric enforcement of simultaneity for the country as a whole:

It is on record that a small religious body once adopted two resolutions as a declaration of its faith. The first was,
Resolved, That the saints should govern the earth. Second,
Resolved, That *we* are the saints.[59]

Resolved or not, some among the railroaders saw the time masters as anything but saints. This opposition was not, as Allen sometimes suggested, based on a reactionary attachment to the removal of God's own noonday sun. Protesters uniformly *accepted* the principle of a railroad-and-telegraph-determined simultaneity. There were, after all, *already* effective time zones carved out by railroads from the Atlantic to the Pacific. A railroad guide in September 1883, for example, showed some forty-seven lines running on New York time while thirty-six took time from Chicago's clocks, while Philadelphia's clocks governed another thirty-three.[60] *Every* train line accepted the conventionalism of time. This fundamental feature of late-nineteenth-century time was so deeply hammered into rails and wires that it was as present to time-zone protesters as it was to boosters. What then did the dissenters dispute? One newspaper article, defending a single national time, targeted the observatories, where astronomers, in charge of far-flung outposts, had grown fat feeding time to regions; local simultaneity had become "the meat upon which these star-gazers feed." Another anti-standard time complaint came from an almanac publisher: sunrise and sunset times would be thrown into a cocked hat by time unification.[61] Most protesters simply disputed the dividing lines of conventional simultaneity, their antizonism grounded in the convenience of local train schedules, or regional, national, or universal time. Very few invoked "God's true time."[62]

With just a week before the General Time Convention was to begin on 11 October 1883, Allen told a dissenting Massachusetts train line that he now had some seventy thousand miles of road lined up behind time reform. As mileage mounted, he treated the opposition with less solicitude.[63] Cities too fell into place. In Boston, the trains agreed to switch to standard time if the Harvard Observatory and its dependent institutions would do so as well. When the time convention delegates at Chicago's Grand Hotel received a crucial telegraphic concession from the Bostonians, it was clear that the reform would pass: "The city, the railroads, and the observatory only await affirmative vote of Convention before fixing date for changing all public time in Boston." Applause echoed through the room.[64] Saluting the inevitable, the Naval Observatory too agreed to the zone system, setting aside its national-scientific aspirations.[65]

When the vote came, railway officials canvassed not by the number of delegates for or against, or even by the number of companies saying yay or nay. Instead voting was by mile of track (appropriately enough). For the reform: 27,781 miles of track within the convention plus 51,260 miles of nonmember rails, yielding 79,041 iron miles in favor of Greenwich-based time zones. A mere 1,714 track miles lay opposed. "Resolved, that we hereby pledge ourselves to run the trains upon our respective roads by the standards agreed upon and to adopt the same when the next schedule goes into effect," 18 November 1883.[66] Allen wanted a system of branched electrically coordinated clocks that would join all the time balls and train clocks and urban clocks together, pinning the whole to Greenwich.[67]

As the seventy-nine thousand miles of railroads hoped, New York agreed to zone simultaneity on 19 October 1883, when Mayor Franklin Edson signed his approval. Edson forwarded the recommendation to the city's aldermen wrapped in a sheaf of railway documents and institutional endorsements. By year's end, conventionality talk about time was as common in New York as Paris; even an American big-city politician could proclaim: "What is called local time is only a mean which differs from astronomical time, but which suits the convenience of the people in any locality because all agree to use it." Since the majority of the community benefits and difference is slight, the mayor reasoned, why not conform to the new standard?[68] "Conventional and arbitrary," chimed the city fathers, time "should be accommodated to that which is most agreeable to the interests of the people whom it is designed to direct." With those words came the resolution that at noon on 18 November 1883, not only the railways but also New York City time would be that of the seventy-fifth meridian, running several minutes west of City Hall.[69]

Time reformation had passed from a myriad of competing simultaneities to a tightly coordinated fact, plucked from telescopes, confirmed by humming iron rails, then wired into metropolitan clocks. Barnard wrote Fleming on 22 October 1883, reporting that the railroads had finally successfully attacked time: "This settles the question forever, for this hemisphere. In Europe, we cannot hope for a similar result in our time. Paris will probably never consent to use Greenwich time, but we need not concern ourselves about that." Perhaps, Barnard mused, the North American triumph of zone simultaneity "[m]ight stir up dilatory governments to take action in regard to the proposed Prime Meridian Conference at Washington."[70]

Time into Space

In the Paris of mid-1884, however, observatory time was not, for a single minute, mainly about trains and city clocks. The Paris Observatory had never been structured as a commercial enterprise. Time-distributing astronomy, at least among the Polytechnicians and their allies, did not have the slightest role (as for many of their British counterparts) in sanctifying empire with natural theology. Conversely, even among Anglo-American metric boosters, the charms of the meter tended to be charms of facilitated, international commerce. For French savants including Poincaré, his teachers, and colleagues, exchange advantages might well be apparent. But there was far more to decimalization, unification, and rationalization than that. These were longstanding Enlightenment ideals, ideals that joined French aspirations to restore national dignity after the "disaster" of 1870 to the establishment of the secular, progressive rationality that they held to be the hallmark modernism of the Third Republic itself.

For Poincaré, progressive and technical goals were united in the Bureau of Longitude, one of the great centers of enlightened science since the Revolution. By 1884, the Bureau's principal activity was using observatory-based electrical time to map the world electrically. Indeed, for the whole period from the mid-1860s to the 1890s, France, Britain, and the United States raced to establish simultaneity over a sprawling network of undersea telegraph cables to fix longitudes and redraw the global map. This race for symbolic map-possession contributed to the explosive atmosphere surrounding the prime meridian showdown slated for October 1884 in Washington, D.C. But the stakes involved in setting the electric worldmap were higher still. In 1898, when Poincaré argued in "The Measure of Time" that simultaneity was a convention, he had been for over five years a "member" (one of the handful of guiding figures) at the Bureau, serving in that capacity from 4 January 1893 until his death in 1912. In September 1899, about a year and a half after he published "The Measure of Time," Poincaré was elected president of that institution, a post he rose to again in both 1909 and 1910. No figurehead, Poincaré issued reports, led commissions, and oversaw longitude operations during some of his most productive years.

The Bureau of Longitude? One might imagine such a calculational bureaucracy to be of little interest when set against the higher reaches of Poincaré's mathematical physics or his inquiries into non-Euclidean geometries, the stability of the Solar System, or daring philosophical accounts that put conventions in place of absolute truth. But the Bureau's task is central to our story. Rightly understood, the vast theory machine it ran will alter our understanding of a turning point in Poincaré's reconceptualization of time.

On 7 April 1884, the astronomer Hervé Faye stood before the Academy of Sciences in Paris to read a report by Lieutenant Octave de Bernardières that raised a crucial issue. De Bernardières was one of the first of a new type of naval officer trained not only for his ocean-going career but also brought up, so to speak, by the astronomers at Montsouris. He so excelled in his astronomy that he was soon co-authoring a massive 336-page report on the very unoceanic longitudinal difference between Berlin and Paris. But to the Academy of Sciences, de Bernardières reported that the precision of longitude determination had increased dramatically over the previous years, sometimes thanks to the transport of accurate clocks and sometimes by astronomical means. In 1867 these older techniques fell by the way as the first cable spanned the Atlantic Ocean, closing the longitudinal gap between England (Greenwich Observatory) and the United States (Naval Observatory, Washington). Throughout the 1870s and 1880s, the new technology undergirded all long-distance longitude work as telegraphs coupled to submarine cables crisscrossed all the world's oceans.[71]

In Britain, especially, factories churned out prodigious quantities of cable. First, a thick copper conductor was insulated by commercial "gutta"—a newly developed mixture of gum, gutta-percha, resin, and water. Then the manufacturers wound jute yarn to provide a cushion between the gutta-coated cable and a ring of thick iron wires to protect the copper core from breaking. More jute bound these iron wires together; then more wires with more yarn (at least near the dangerous rocky shore), and a final waterproof outer sheathing of the Malayan rubberlike gutta-percha. Steam-powered ships carried the mile-long sections out to sea, where onboard cable masters tied them together to span thousands of miles and sometimes hauled them out again when they (all too often) were split by sea life, icebergs, volcanoes, anchors, or sharp rocks.[72]

As de Bernardières knew well, the French had long been frustrated by the discrepancies in map making that resulted from long ocean voyages (which rendered chronometers unreliable). The French Bureau of Longitudes was directed in 1866 to fix clear secondary locations around the globe. It duly launched six parties to the various corners of the world. Their assignment was to use the

position of the moon against the background of the stars to determine the local longitudes of sites in North and South America,
Africa, China, Japan, along with others in the Pacific and Indian
ocean islands. Such mapping demanded a monumental effort, and
the French government withheld neither expense nor effort. In
principle, the idea was simple: if astronomers in two parts of the
world could both pinpoint the time at which the position of the
moon reached its highest point on the celestial sphere, then those
observations would be simultaneous. The navigator would look up
on a chart what time that was back home and observe what time it
was locally. The difference gave the longitudinal difference between
the two points. A difference of six hours? –90 degrees of longitude.

But moon culminations were notoriously elusive. Even the most
skilled astronomers seemed unable to fix the moon's zenith against
the stars without significant errors and that was a huge problem.
Here is why. The earth spins on its axis once per day, making the
stars appear to rotate every twenty-four hours. The moon travels
much more slowly against the background of the stars, about once
every thirty days. So in the time the moon crosses a given angle, the
stars have moved thirty times that angle, thereby multiplying any
error by thirty. Piloting a ship through distant shoals with that kind
of uncertainty killed sailors.[73]

Because of the difficulty of "shooting the moon," astronomer-
surveyors hunting their longitude reached for other, more easily
measured, heavenly events. Navigators had been using total eclipses
for centuries; Columbus used one to fix his longitude on his transatlantic explorations. So when American map makers wanted to fix
the longitudinal relation between Washington and Greenwich,
observing the moon-darkened sun seemed promising. The United
States sent a steamer to Labrador on 18 July 1860, hoping to compare the results to that same event as viewed from Spain.

Another longstanding practice of the astronomers was to watch
the moon as it arced across the night sky until a star winked out

behind it. That moment of disappearance (occultation) could also be exploited to set simultaneity between distant points. Measure the time of occultation locally, look on a chart (or wait for a report by post) to find what time that same event was observed in Greenwich or Paris, then subtract one from the other. So astronomers both in the United States and Europe watched and measured with enormous care as four stars of the Pleiades dipped under and then out from behind the moon. An occultation of the planet Venus took place on 24 April 1860; astronomers huddled in wait for it in the Fredericton, New Brunswick, and Liverpool observatories. Finding longitude was part of the job description of French, British, and American astronomers, and they deployed every conceivable method to find it. But the quest to link the United States to the "well determined European observatories" remained unfulfilled. Every expedition produced a new number.[74]

Mapping offered both symbolic and practical mastery over space. In the great land grab of the mid-nineteenth century, fixing positions was critical for trade, for military conquest, for laying railroads.[75] When the United States plunged into civil war, the Coast Survey became a strategic asset. Congress had long before demanded that the surveys cover rivers as needed by commerce or defense. Now map makers hoped to deliver on that assignment, working closely with Union admirals in North Carolina and on the Mississippi. At first, the telegraphic surveyors, including George Dean, set about reducing data already in hand to provide precise positions of key southern military sites, determining longitude differences between them.[76] Watching, measuring, calculating: the surveyors plotted Rebel emplacements around Charleston and Savannah, while a small group of reconnaissance topographers joined General Sherman on his march from Savannah to Goldsboro, Georgia. When the war ended, the surveyors began plotting a future using their wartime telegraphic surveys. They had new and better measurements of almost every major town from Calais,

Maine, at the northeastern edge of the United States, down to New Orleans. Benjamin Gould (who headed the survey's telegraphic longitude effort) and his team looked eastward to fill the one critical gap—New York City to Washington—that remained in the quest for a complete electrical map of the United States east of the Mississippi.[77]

The surveyors looked to the ocean. They did so in some desperation, because no matter how hard the survey had tried, convergence on a longitude difference between Europe and the United States continued to prove infuriatingly elusive. They reviewed moon culminations, restudied data on the occultations of stars and planets, and pored over old chronometer results. But reanalyzing old data was simply not enough: "The discordance of results, which individually would have appeared entitled to full reliance, is thus seen to exceed four seconds; the most recent determinations, and those which would be most relied upon, being among the most discordant. No amount of labor, effort, or expense had been spared by the Coast Survey for its chronometric expeditions, inasmuch as the most accurate possible determination of the transatlantic longitude was specially required by law." In addition, the latest chronometer studies differed from the best astronomical studies by an embarrassing and irreducible time difference of three and a half seconds.[78] No one had any idea which result to believe.

Only underwater electrical telegraph cables could break the impasse. Beginning in August 1857, missions had struggled against the North Atlantic to lay a telegraph line. Cables broke again and again; in June 1858, the fleet once more launched from Plymouth, England, with the vast cable. Just three days out at sea, a gale tore at the ships ceaselessly for nine days. Despite significant damage to one ship (and one sailor so terrified that he lost his reason), cable-laying continued. On 6 August 1858, the first signals finally passed through the cable, followed by a cable break and the cessation of submarine cabling during the Civil War. Pulling out from Valentia

Island in southwestern Ireland in July 1865, the *Great Eastern*, a mammoth ship five times larger than any other, began hauling cable toward Newfoundland. Twelve hundred miles later, both the cable and the ship's lifting gear sank into the sea. The mission was aborted.[79] In 1866 another crew set out, this time aiming to run a much-improved cable from the tiny fishing hamlet at Heart's Content, Newfoundland (on the eastern side of Trinity Bay, about ninety miles from St. John's) to Valentia. This time: success. Communication started on 27 July 1866, and the surveyors immediately began sending time signals. Gould dispatched his longitude team in small groups to staff the dilapidated way stations up the East Coast. To inspect the telegraph line from Calais to Newfoundland, the expedition chartered a schooner, plying from Nova Scotia's Cape Breton Island to their destination at Heart's Content, Newfoundland. Every mile, repeaters (relaying the signal) needed inspection; they had to crash-train dozens of isolated telegraphers.[80]

Gould himself set sail for Liverpool and London on 12 September 1866 by the British mail steamer *Asia*, first to confer with the British cable company officers and then to bring his measuring instruments to the Irish terminus. Bumping through the Irish countryside on a jury-rigged spring-cushioned cart, the astronomers hauled their precarious tower of crates forty-two miles, then ferried their goods to Valentia across the straits from Killarney. Conditions at cable's end were discouraging. The British company refused the Americans permission to pull land lines into electrical contact with the undersea cable for fear a lightning strike would damage their fragile leash on the New World. That meant the Americans had to set their observatory adjacent to the telegraph company building at Foilhommerum Bay, "remote," as the time men put it, "from any other dwelling-house except the unattractive cabins of the peasantry." On this rugged terrain, the time men banged together their makeshift observatory, 11' x 23', bolted to six heavy stones buried in earth, protected from southwest gales by the adjacent telegraph house, and shielded by rising ground from northwest weather. The larger room was their observatory; the smaller, on the eastern end, became their dwelling. Their lab was simple: a rigid mount for the transit instrument, nooks for clock and chronograph; a spot for the relay magnet that would pass the signal along toward Greenwich, the Morse register, and a recording table.[81]

From what Gould dubbed Valentia's "peculiarly unastronomical sky" came rain. Bucketsfull. Clouds allowed but one or two glances of the noontime sun. When the sun did poke through, the astronomers, ever vigilant, fixed their meridian. Finally, on 14 October 1866 at 3:00 A.M., the Americans glimpsed a few stars through the haze and took transit readings. Locals reported that for the eight weeks prior to the surveyors' arrival it had rained at Valentia every day without exception. During the seven weeks that the survey crew lived and worked in their Irish shack by the sea, they saw four days without rain and only a single clear night. "The observations were, in general, made during the intervals of showers; and it was an event of frequent occurrence for the observer to be disturbed by a copious fall of rain while actually engaged in noting the transit of a star." Time sentries at the other end of the line in Newfoundland had it worse. Telegraphic Surveyor George Dean spied nothing, not a single glimpse of the sun, moon, or stars.

Here was Victorian high technology on the ragged edge of Britain, powering their transoceanic signal with a battery composed of a percussion gun cap with a morsel of zinc and a drop of acidulated water. Weather gods permitting, the Irish station would cable "GOULD" in Morse code; Newfoundland would respond "DEAN," followed by time signals—half-second pulses punctuated by five-second intervals. At both ends of the cable, teams hovered over instruments. After crossing the Atlantic, the signal was too weak to drive the drum recorder, so they used the mirror galvanometer, a vastly more sensitive device invented by British physicist William Thomson. A delicately suspended mirror with a tiny magnet glued to its back reflected light from a kerosene lamp. Nearby was a coil wired to the undersea cable. When a signal current coursed through the cable, the coil became an electromagnet, gently twisting the little permanent magnet with its attached mirror, which shifted the reflected light of the kerosene lamp against a mounted sheet of white paper. Even the weakest of transoceanic time signals became visible. Anticipating the signal, observers would focus the bright light of their kerosene lamp on the mirror. And wait, hour after hour, in the cold, damp night, hoping that an electric current crossing under 4,320 miles of ocean would dance a tiny spot of reflected light across a soggy paper sheet.

The mobile astronomers strictly followed their disciplined hard-won procedures, pried from longitude campaigns that both Gould and Dean had run during the Civil War. Gould's team received its first signal on 24 October 1866; in the weeks that followed, Gould and his telegraphers eked out four more exchanges in tiny windows of astronomical weather. An ocean from Gould, in Newfoundland, Dean's struggle to relay the signal down to Boston was much less successful. As the cable snaked along the eleven hundred rough miles from Heart's Content to the entry point into the United States at Calais, Maine, communication broke down "day after day, and week after week." Nothing worked. Suddenly, on 11 December, a sharp frost gripped the defective land line coming into Calais. All at once ice admirably insulated the wire and the pulse shot like lightning from Heart's Content to Calais. Just before New Year's Day 1867, the Newfoundland team steamed into Boston Harbor, having bagged the longitude difference between Harvard and Greenwich observatories.[82]

With the Atlantic wired for simultaneity, the tempo of electric map-making only increased. Immediately following the British-American collaboration, the French hauled a line from Brest through St. Pierre off Newfoundland to Duxbury, Massachusetts. American astronomer-surveyor Dean and his peripatetic team of surveyors deployed again, almost immediately, to begin planning with the French naval authorities for a time check. Once the Brest-Paris line was secure, the surveyors had their dream, a series of triangles that would serve as a double check of their longitude results. For example, the longitude differences ought to (and did) add to zero if one went from Brest to Paris to Greenwich and back to Brest:

(Brest-Paris) plus (Paris-Greenwich) plus (Greenwich-Brest) = 0.[83]

The French Naval Lieutenant de Bernardières was all too aware that his American and British counterparts were on the longitudinal march. In the spring of 1873, at his superiors' instigation, U.S. Naval Lieutenant Commander Francis Green had begun sending suboceanic time signals to map the West Indies and Central America. Traveling the seas in the side-wheel steamer *The Gettysburg*, Green's team brought precision to the longitude of Panama, Cuba, Jamaica, Puerto Rico, and many other islands.[84] On his return in 1877, the authorities set the lieutenant commander to work again, this time to exploit the newly laid telegraphic cables crossing the Atlantic: London to Lisbon to Recife in northeast Brazil. For the first time the American Navy could ply its hydrographic trade all along the eastern coast of South America from Para to Buenos Aires. At last, the Navy boasted, it would settle the much-disputed locations of places like Fort Villegagnon (in the Bay of Rio de Janeiro), where previous longitude men had clashed over an astonishing thirty-second time discrepancy, an eight-mile uncertainty over where one's ship would hit the eastern edge of South America.

With the help of the French, who shot their Parisian signals down through the wires to Lisbon, the Portugese too joined the Greenwich-based map-making scheme.[85] At Porto Grande off the Brazilian coast, the longitude team halted, waiting for the "sick season" to subside before they dared land at Pernambuco: "A less interesting place than Porto Grande to pass two or three weeks could not be found," groused the Americans. "The island of St. Vincent is merely a heap of cinders."[86] Finally, the astronomers hopped a royal mail-steamer into Pernambuco and set the machine into motion: a pulse from Greenwich Observatory to Land's End, down through 828 miles of undersea cable, out from the instrument room of the Eastern Telegraph Company near the lighthouse at Carcavellos, then back out from the Royal Observatory of Lisbon across the Atlantic, into the grounds of the Brazilian Naval Arsenal, up insulated wire over the arsenal walls, along the roof of the port captain's office, through the esplanade, by balconies, and down to one of Thomson's mirror galvanometers, where it would sway its beam of white light.

Subtle, evanescent, and unreliable. Yet when that electrical pulse finally reached Rio in July 1878, it was of royal significance. Brazil's Emperor Pedro II had been in the United States during 1876, then launched a great European tour, visiting Troy with Heinrich Schliemann, receiving communion at the Holy Sepulcher in Jerusalem, and reveling with Queen Victoria's daughter and her husband in Vienna. In Paris the Academy of Sciences inducted Pedro as a foreign associate; Victor Hugo received him in celebration of literature. Back in London, he lunched with the Queen at Windsor Castle. Given this public life, and apparently an equally entertaining private one, Pedro returned somewhat reluctantly to Rio. Nothing, however, could keep His Majesty away from Lieutenant Commander Green's ramshackle observatory to witness the electrical arrival of European time.[87]

Today the word "observatory" may bring to mind a romantic image: a glistening hemisphere perched on the craggy peak of a mountain; the astronomer, sliding the sky slit open, swivels his great brass telescope to view the heavens, while white-coated assistants shuffle quietly at his side. Not for Green and his U.S. Naval team, who would come to town (such as it was, say, in Porto Grande) and set up shop. Finding a location with a good view of the sky near the telegraphic office was obviously key. They would then scrape a meridian (north) line onto the ground and build a cement and brick pier. On top the astroexplorers set a small slab of marble that they carried with them and, on top of that, their precious brass transit instrument, through which they could watch stars cross a spider's thread that marked the meridian line. It would take two sailor-astronomers about an hour to assemble their portable wood observatory (8' x 8'), with a canvas top in case of rain. Into this tiny station Green crammed clocks, telegraph keys, and either a recording drum or a Thomson galvanometer to display the incoming signal. When the observer needed a telegrapher at his side, the room was, one could say, cozy.

Having entertained Pedro II with the electromapping of Rio, the American Navy sailed on. Early in 1883, Washington ordered the team to the west coast of South America now that cables joined it too to the telegraphic system. This new expedition followed cables running from Galveston, Texas, through yellow-fever-infested Vera Cruz in southeastern Mexico. With Vera Cruz on the electric worldmap, they sailed down the western coast of South America, hoping to exploit the cable binding Salina Cruz in southeastern Mexico to Valparaiso, Chile. The team faced down quarantines at New Orleans and Galveston, only to arrive in Peru in the midst of the Chilean military occupation.

Thanks to the intervention of the American minister to Peru, Mr. S. L. Phelps, the occupying admiral of the Chilean army promised to assist the time bearers. From Lima, the navy men headed south on 13 October 1883, tying Valparaiso to Lima, and by early 1884 they were banging together their wooden observatory in Paita, Peru. Pounding rain drenched the desiccated town for the first time in seven years: "the soil of the place, ordinarily arid and dusty, was converted into a sickening and fetid mud . . . the whole town . . . rendered almost uninhabitable."[88] Furthermore, "operations in Peru," the officers reported, "were at points occupied as military posts by

Figure 3.9 Portable Observatory: Bahia, Brazil. American Naval officers determining simultaneity — and therefore longitudinal differences — in order to map more precisely the Americas. Longitude expeditions by French, British, and American teams were part scientific, part military, and part exploration, the "astronomical observatories" often no more than portable wooden shacks accompanied by a few pieces of electrical telegraph equipment, some magnets, mirrors, and a surveyor's telescope. They came at a cost: numerous surveyors died of disease and accident while trying to pin electrical simultaneity to the shores of the Americas, Asia, and Africa. SOURCE: GREEN, REPORT ON TELEGRAPHIC DETERMINATION (1877), FRONTISPIECE.

an invading army, and the observers had to contend with the dilatoriness, and in the case of the military commandant at Arica, the indifference and stupidity, of officials," not to speak of custom delays, fog, and broken cables. Yet on 5 April 1884, the American officers set sail for New York with their South American longitudes.[89]

In tiny, far-flung observatories, the American, British, and French sentries of time took their readings from the skies and matched them against the pulses from their cables. Stars told each station its local time, cables whispered to them of time somewhere else. It was devilishly hard work, demanding precision results, strong cables, and orchestrated teamwork. One of de Bernardières' French collaborators perched his station near the ruins of the city of Chorillos, just south of Lima. Another waited by his telegraph in the observatory of the naval school in Buenos Aires. Linking up the various wires, de Bernadières began signaling electrically from Valparaiso to Panama. All along the route, relay men struggled to keep the signal moving without a break: the tiny current tickling the mirror; reflected light flickering. As soon as he saw the light spot move, the operator would relaunch the signal. Starting on 18 January 1883, while the Americans were still readying crates in Washington for their expedition along the coast of Peru and Argentina, the French team seized three excellent astronomical and electrical evenings. Just after their last measurement, the submarine cable into Panama snapped, severing communication between their minuscule observatory and Paris. But they had results: the longitude of the flagpole over the stock market of Valparaiso stood 4 hours, 55 minutes, and 54.11 seconds earlier than the Montsouris time room thousands of miles to the east.

It was just after this intense round of competition and collaboration with the Americans that, on 7 April 1884, de Bernardières issued his plea to his academician colleagues in Paris. His report told the scientists that there were new possibilities for French surveying: two great oceans had just been linked across South America by telegraph, with cables snaking their way at 12,000 feet across the formidable Andes. As de Bernardières put it, the Paris Bureau of Longitude should now imagine creating an "immense geodesic network that would encompass the entire globe, fixing precisely its form and dimensions."[90]

The French Navy backed the plan, lending officers, sailors, and matériel. De Bernardières would install himself at Santiago or nearby Valparaiso, and others would tell time to the north in Lima and Panama and across the continent in Buenos Aires. Just a few months earlier, the American company Central and South American Cable had laid their undersea line from Lima to Panama. Adding to an already established line stretching from Panama through the Greater Antilles, North America to Europe, de Bernardières and his colleagues had a filamentary link all the way back to Paris. Since other lines also went from Valparaiso through Buenos Aires, the Cape Verde Islands and eventually to Europe, the whole constituted a vast circuit of some 20,000 miles of gutta-percha-covered copper, allowing the two routes to check one another.

Out of Green's American mission and de Bernardières's French one came an enormous, world-clasping polygon of simultaneity with vertices in Paris, Greenwich, Washington, Panama, Valparaiso, Buenos Aires, Rio de Janeiro, and Lisbon. Amazingly enough, reckoned in two different directions, this irregular octogon came within 150 yards of closing. From the polygon, spokes jutted out to Asia where the Americans were electrically grasping the contours of India, while the French, from a fragile bamboo hut, a few pieces of electrical equipment, some astronomical gear, and a cable-end, began their electrical pinpointing of Haiphong.[91]

This French world net of cables and time signals shot east, west, north, and south from Paris. Astronomers shaped much of its character, and in turn, the longitude machine transformed their observatories. Take, for example, the fourth volume of the *Annales du Bureau des Longitudes*, which in 1890 opened with an account of the Bordeaux Observatory. Whatever research might yield, the report made plain, the observatory was built neither to scout new phenomena nor to place the human form in a cosmic context. No, the city of Bordeaux, like so many others, had built an observatory to set ship's chronometers so they could determine their longitude

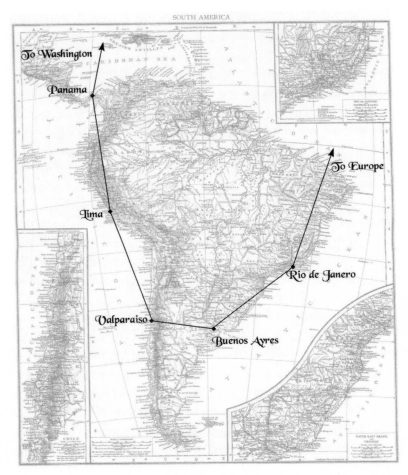

Figure 3.10 Cabled Time in South America. *One of the major French longitude explorers, de Bernardières raced against the Americans to complete an electric-time-based cartographic net tied back to the Paris Observatory. To his colleagues back at the Academy of Sciences, the Frenchman urged the creation of the polygon sketched here: "an immense geodesic network that would encompass the entire globe."* SOURCE: MAP MODIFIED FROM *THE TIMES ATLAS*, LONDON, OFFICE OF THE TIMES, 1986.

at sea.[92] This meant that the Observatory's very first task was to establish its own longitude, which it completed by using telegraphic time signals with Paris on 19 November 1881, locating itself 11 minutes and 26.444 seconds west of the capital—plus or minus eight-thousandths of a second.

City by city, country by country, the French Bureau of Longitude extended its network of longitudinal fixed points, first nationally, running telegraph wires from Paris to the distant cities of France, and then, through an array of undersea cables, to the distant colonies. By 1880, ninety thousand miles of mostly British cable lay on the ocean floor, a ninety-million-pound machine binding every inhabited continent, cutting across to Japan, New Zealand, India, through the West Indies, the East Indies, and the Aegean. Competing for colonies, for news, for shipping, for prestige, inevitably the major powers clashed over telegraphic networks. For through copper circuits flowed time, and through time the partition of the worldmap in an age of empires.

As maps merged, a consensus emerged that a universally acknowledged first meridian was needed. That origin arc would set one nation's capital at the symbolic center of every global map. Every clock and every longitude measurement in the world would refer back to the dead center of one country's transit instrument. Though the immediate struggle was over a symbolic centrality rather than control of land, no one mistook its importance. High noon for the diplomats and scientists was to be 1 October 1884 in Washington, D.C., in the Diplomatic Hall of the American State Department.

Battle over Neutrality

Two years prior to the meeting, in August 1882, President Chester A. Arthur and the American Congress had approved a resolution calling for an international conference to be assembled in Washington for the purpose of setting a unique and universal prime meridian.[93] While American politicians came to consensus, a group of scientific delegates gathered for an international geodesic conference convened in Rome in October 1883. Swiss astronomer Adolphe Hirsch reported on their deliberations, starting with their celebration of the network of telegraphic longitude fixes that had finally specified reference points across the European continent. With many countries producing maps with their own capitals as the zero longitude point, a new question had arisen: Could all these reference points be referred back to one single and unique prime meridian line? Now, Hirsch recorded, it was time to set aside ideals of a science isolated from the world and instead to make a contribution that, in a far deeper way, joined science and the wider practical domain. Great nations had an opportunity to help navigation, cartography, geography, meteorology, rail transport, and telegraphic communication. It was time, once and for all, to choose a universal prime meridian. Just because the earth was a sphere, Hirsch insisted, *there was no natural prime meridian.* That is, a prime latitude—the equator—was picked out naturally by the spin of the earth. But "nature" had selected no prime meridian of longitude. Even the magnetic north pole could not be used to single out any particular longitude as prime, as the magnetic north wandered about over time. No, the choice of a prime meridian was necessarily arbitrary and therefore subject to reasons "purely practical and conventional [*conventionelle*]."

Here was a theme the experts repeated over and over again: the exchange of ideas, products, and peoples demanded new international institutions even while respecting the individuality of nations. Global conventions must prevail over local ones. Unions of post and telegraphs now embraced the entire world. The Convention of the Meter already united the majority of civilized countries, there were conventions of electrical standards; conventions to protect intellectual, artistic, and industrial property; conventions to protect combatants in conflict. Even their own geodesic association testified that a purely scientific goal—the establishment of the exact shape of the earth—could propel such coordination among nations. Hirsch insisted that it was time to find a practical solution to the unification of longitude, and the conference therefore recommended that Greenwich become that prime meridian, with a date change at the antimeridian on the opposite side of the earth.[94]

A year after Rome, the delegates to the Washington Conference joined U.S. Secretary of State Frederick T. Frelinghuysen at the State Department. With a rigorous record of the conference prepared and approved as it progressed, the statesmen and scientists joined the battle over the longitudinal zero point. Poincaré of course read these proceedings; he even cited them verbatim in his later articles. He, along with many others who studied the proceedings, witnessed the inextricable mix of politics, philosophy, astronomy, and surveying.

After a formal welcome by the secretary of state in which he pointedly renounced establishing the prime meridian in the United States, Lewis M. Rutherfurd of the American delegation shot the starting gun by proposing, as had the Rome Conference, that the standard meridian pass through the center of the transit instrument at Greenwich Observatory. French Minister Plenipotentiary and Consul-General to Canada Albert Lefaivre immediately cautioned against hasty action, belittling the Rome decisions as the product of mere "experts." Here in Washington, the loftier vantage point of politicians ought to prevail: "It is, moreover, our privilege to be philosophers and cosmopolitans, and to contemplate the interests of mankind not only for the present, but for the most distant future."[95] Only such a distanced view would allow consideration of principles. Backing down, the Americans proposed that the conference merely accept the idea of a single prime meridian.

Such backtracking offended the British. Captain Sir F.J.O. Evans of the Royal Navy protested that Rome had already narrowed the question: the prime meridian should bisect a great observatory, not a mountain, a strait, or a monumental building. Science, after all, demands precision, not some vague gesture toward a natural feature of the earth. And the list of such great observatories was short: Paris, Berlin, Greenwich, and Washington.[96] Commander Sampson of the United States Navy concurred, saying that the chosen observatory must have the requisite telegraphic time links to the whole world: "We may then say that, from a purely scientific point of view, any meridian may be taken as the prime meridian." If one demanded convenience and economy there were fewer options, and that choice narrowed considerably if the prime meridian had to pass through a thoroughly wired, governmentally backed observatory. If one recognized that any prime meridian other than the British one would require altering maps for the 70 percent of world shipping that used the Greenwich meridian, there was only one real choice: Greenwich.[97] No doubt encouraged by the twin salvoes of the British and American navies, Rutherfurd rejoined the siege of Paris: "The observatory of Paris stands within the heart of a large and populous city," a city subject to movement of air, tremors of the earth. The Paris Observatory had to be moved. "The only thing which keeps it there is the remembrance of the honorable career of that observatory in times past."[98]

Not so, the French astronomer and delegate Jules Janssen retorted. The Paris Observatory remained vital because it was linked electrically with all the other major observatories, its abilities well documented by a use of telegraphic mapping altogether comparable to Greenwich. What mattered historically was this: Following Ptolemy, Cardinal Richelieu had set the prime meridian on the island of Ferro in the Canary Islands. Since the easternmost point of Ferro rose from the sea some $19°55'3''$ degrees west of Paris, reckonings became hard to calculate. Eighteenth-century French astronomers simplified longitude calculations by decreeing that Ferro's prime meridian would be, statutorily, exactly 20 degrees west of Paris. Since even French astronomers could not physically move the island of Ferro 4'57" east, Janssen admitted that this convention made the prime meridian Paris, disguised.

No doubt Janssen could read the politics as well as anyone else: the likelihood of replacing Greenwich by Paris (disguised as Ferro) as the prime meridian was nil. But Janssen was not about to fold his tent. (This was an astronomer who, during the siege of Paris in 1870, had taken off in violent winds in a balloon for Algeria to observe an eclipse.) It was time to seize loftier ground than Ptolemaic (and now French) Ferro, for the conference was beginning to skid toward the crassest of decisions. "Instead of laying down the great principle that the meridian to be offered to the world as the starting-point for all terrestrial longitudes should have, above all things, an essentially geographical and impersonal character, the question was simply asked, which one of the meridians in use among the different observatories has (if I may be allowed to use the expression) the largest number of clients?"[99]

Clients. The very thought offended a rational (French) sensibility. Janssen hoped that customs and customers would not throw industrial (British) smoke before philosophical (French) principles. There was, he reminded the assembled, the long tradition of French hydrographic engineers; there was the universally respected almanac, the *Connaissance des Temps*. And not least his colleagues ought to remember that it was the French who, in revolutionary times, had stepped past the *pied de Roi* as a unit of length in favor of the rational meter. Rational science should trump royal trade.[100] Janssen opined: "Without doubt, on account of our long and glorious past, of our great publications, of our important hydrographic works, a change of meridian would cause us heavy sacrifices. Nevertheless, if we are approached with offers of self-sacrifice, and thus receive proofs of a sincere desire for the general good, France has given sufficient proofs of her love of progress to make her cooperation certain." A reasonable agreement, Janssen concluded, cannot protect only one contracting party.[101] In other words, put the longitudinal center of the world anyplace neutral. Anyplace but Greenwich.

Come, come, the British astronomer John Couch Adams harrumphed. It was Adams whose work had already been twice mobilized in skirmishes with the French, once over the question of who had discovered Neptune and again when he contradicted the results of the great Laplace on the motion of the moon. We are not belligerents, he now insisted, we are all neutral, as in all matters scientific. We are not dividing territory as after a war but in a friendly way representing friendly nations. What will provide the greatest convenience to the world? And we should provide that convenience without the legal fictions of a displacement from some other observatory (Paris disguised as Ferro), but rather calling things "by their right names." Practical, hard-headed decision making ought to be the order of the day: "It was quite clear that if all the Delegates here present were guided by merely sentimental considerations, or by considerations of *amour propre*, the Conference would never arrive at any conclusion." To the charge of French vanity, Janssen answered by accusing the British of vulgar self-indulgence:

> We consider that a reform which consists in giving to a geographical question one of the worst solutions possible, simply on the ground of practical convenience, that is to say, the advantage to yourselves and those you represent, of having nothing to change, either in your maps, customs, or traditions—such a solution, I say, can have no future before it, and we refuse to take part in it.[102]

Loyal to the practical and commercial, the Americans sided with the British. What is neutral? demanded Cleveland Abbe. Historical, geographical, scientific, or arithmetical neutrality? True, France gave us neutral weights and measures, but those measures have an arbitrariness to them by virtue of the standard weights and measures on which they rest. "Neutral" system of longitude is "a myth, a fancy, a piece of poetry," unless you can tell precisely how to do it.[103]

Janssen retorted: A neutral point would have two advantages, geographical and moral. Choosing the Bering Strait would remove the prime meridian from all centers of population for the statutory change of date, it would neatly slice the globe into the Old World and the New. Or, if not the Bering Strait, then another remarkable physical point; laying the zero point of time and longitude through the Azores would cost less than the Bering Strait because telegraphic cables already ran near. In either the Azores or the Bering Strait, one would define the zero of longitude from existing telegraph-connected observatories, not the center of the strait itself. (Janssen had just noted the excellent electrical connectivity of the Paris Observatory.) No doubt looking pointedly to Adams, Janssen urged his colleagues to bear in mind the "lively discussion" raised by the English and French press over the discovery of Neptune (both sides had claimed astronomical priority). Delving deeper into the past, Janssen saw those same Continental-British tensions in seventeenth-century battles over the calculus between defenders of Newton and those of Leibniz: "The love of glory is one of the noblest motives of men; we must bow before it, but we must also be careful not to permit it to produce bad fruits."[104]

Outnumbered, Janssen soldiered on: economic reasons might favor Greenwich, Washington, Paris, Berlin, Pulkovo, Vienna, or Rome, but such choices would necessarily be artificial. "Whatever we may do, the common prime meridian will always be a crown to which there will be a hundred pretenders. Let us place the crown on the brow of science, and all will bow before it." Yes, one Anglo-Saxon responded, but *any* chosen place belongs to a country. Not at all, shot back Janssen; the equator is neutral, yet it traverses nations. England's General Strachey protested against the distinction between longitude for geography and for astronomy—"longitude is longitude." It most certainly is not, Janssen seethed. Longitude depends like any measure on context. "Is not a weighing necessary to determine a chemical equivalent of an entirely different kind from that of a commercial weighing? Yet it is still a weight."[105]

Cleveland Abbe, who had fought so hard with the American Metrological Society for unified time, doubted the naturalness of "neutral." What if Russia were to reconquer the country on this side of the Bering Strait? What if America purchased half of Siberia? "That point [in the midst of the Bering Strait] is not cosmopolitan." Only stars over the earth, something above human considerations, could possibly be deemed neutral.[106]

Just so, interjected Sandford Fleming. Habituated to arguments in terms of the transport of goods and people, he viewed the French proposal for "a neutral meridian [as] excellent in theory, but I fear . . . entirely beyond the domain of practicability." Then, shipping manifest in hand, Fleming recited numbers from a compilation of ships and transport tonnage that had been guided by the various prime meridians: Greenwich, 72 percent of tonnage; followed by Paris, 8 percent, with the rest of the world making up the remainder. Perhaps as a sop to his French colleagues, Fleming offered to set the prime meridian exactly 180 degrees from Greenwich, in the midst of the "uninhabited" part of the Pacific.[107] Such a move would, he added, keep astronomical events practically unchanged on the Greenwich-based charts, inverting midnight and noon, 2:00 A.M. for 2:00 P.M., and so on. As for diplomacy, flipping the longitudinal line across the globe from London's suburb fooled no one, least of all the French, who knew all about proxy prime meridians from their long use of Ferro. On a vote for a neutral meridian, the French plumped 'for,' joined only by Brazil and San Domingo. All twenty-one other nations voted against.

Speaking for France, Lefaivre morosely assessed the debate as void of astronomy, geodesy, or navigation. Reminded of tonnages in the context of Anglo-American complacency, he allowed "the only merit of the Greenwich meridian . . . is that there are grouped around it, interests to be respected, I will acknowledge it willingly, by their magnitude, their energy, and their power of increasing, but entirely devoid of any claim on the impartial solicitude of science." No reason, neutrality, or impartiality—only commerce pure and simple. Lefaivre conceded that the Empire had won by commercial prowess, but on no other grounds:

> Well, gentlemen, if we weigh these reasons—[the] only ones that at present militate for the Greenwich meridian—is it not evident that these are material superiorities, commercial preponderances that are going to influence your choice? Science appears here only as the humble vassal of the powers of the day to consecrate and crown their success. But, gentlemen, nothing is so transitory and fugitive as power and riches.

All empires have fallen, and this one too shall pass. Do not enchain science and subordinate science. And will there be reward for the abandonment of our French meridian? Would America and Britain deign to take up the metric system? No. "We are simply invited to sacrifice traditions dear to our navy, to national science, by adding to that immolation pecuniary sacrifices." Sir William Thomson, the senior scientist at the gathering, confirmed Janssen in all but his low opinion: this was a "business arrangement," not a scientific question. The question was soon called for the adoption of "the meridian passing through the transit instrument at the Observatory of Greenwich as the initial meridian of longitude." San Domingo voted against the sanctification of Greenwich; Brazil and France abstained. Twenty-one nations supported it.[108]

Other issues racked the world longitude system, disputes that only made sense in a world wired for electrical time. Indeed, the whole conference could be seen as an extended conflict over the newly wired world of electrified time distribution. United States delegate W. F. Allen, fresh from his railroad time victory, offered the conference the logical asymptote of all these developments: "It would . . . be one of the possibilities of the powers of electricity that the pendulum of a single centrally located clock, beating seconds, could regulate the local time-reckoning of every city on the face of the earth."[109]

Against this ultimate time fantasy lay a universe of local customs; even the seemingly innocent question, When should the day begin? was vexed. Some favored commencing the day at the antimeridian of Rome to facilitate computation of ancient dates by the Gregorian calendar. Astronomers wanted the day's start at noon to avoid splitting the night into two dates. Meanwhile, the Turkish delegate noted that the Ottoman Empire recognized midnight-to-midnight days (*heure à la franque*) but also reckoned from the bisection of the rising sun by the horizon (*heure à la turque*): "Reasons of a national and religious character prevent us . . . from abandoning this mode of counting our time."[110]

Important as these cross-cultural synchronizations were, no one at the conference doubted that the struggle was between Paris and Greenwich. Defeated in resolution after resolution, the French held one last hope: the decimalization of time. Their ambition for a rationale, scientific measure of time, analogous to the meter, had deep roots in Paris. In Year II of the French Revolution, the *Convention* struggled to institute a decimal system of time keeping with "decades" of ten days (rather than weeks), days partitioned into tenhour units, and right angles split into 100, not 90 parts. A few revolutionary clocks still survive; one (figure 3.11) shows an equilateral tricolor triangle signifying that, under the new liberty, months would be divided in equal parts, with the vertices standing for days of rest. Among scientists, few took up the new system, but Laplace did in his epochal *Celestial Mechanics*, and sectors of the government even tried to impose the system. But faced with massive public resistance, Napoleon killed time decimalization in a deal cut with the Roman Catholic Church.[111]

Janssen took the floor one last time to plead for the global instau-

Figure 3.11 French Revolutionary Clock (circa 1793). The equilateral tri-angle signified that, under the new liberty, months would be divided into equal parts with the vertices standing for days of rest. SOURCE: ASSOCIATION FRANÇAISE DES AMATEURS D'HORLOGERIE ANCIENNE, *REVUE DE L'ASSOCIATION FRANÇAISE*, VOL. XX (1989), P. 211.

ration of this long-deferred decimal time and division of the circle. His hope—put as a resolution—was that the decimal system, now entering the mainstream of European trade and manufacturing, might finally be extended to time. Confronted with the same public opposition faced by his revolutionary forebears, Janssen reassured his colleagues: "It is feared that we want to destroy habits fixed for centuries, and upset established usages." But no such fears were justified. "If we failed at the time of the Revolution, it is because we put forward a reform which was not limited to the domain of science, but which did violence to the habits of daily life." This time it would not mandate a change for the people of the world, but only be imposed where it was of use.

At every stage of the story in France, from the meter to time and longitude, conventions of space and time were powerfully bound to the legacy of the *Convention*. But for all the delegates, and indeed for the larger metrological community behind them, settling on conventions of space and time was never just about precision maps or intersecting rails. To treat the Washington clashes as a step on an inevitable march toward standardized rationality is to miss the fluctuating, contingent, opalescent character of synchronization. It is to miss the constant crossing and recrossing of the pragmatic with the philosophical, the abstract with the concrete.

After the French rout at the prime meridian, delegates looked for a way to accommodate the French on decimalized time without endorsing either its Enlightenment ambitions or any practical plan for its implementation. In the end, the conference merely expressed "the hope" that further technical studies on decimalizing time would be resumed.[112] It was not much for the French delegates to carry home. To advance the cause beyond that minor, conciliatory vote, the French needed someone with stellar scientific credentials and engineering-administrative clout to champion this radical reformation of time conventions. For almost a decade the issue simmered uneasily in French technical circles. Enter Henri Poincaré.

Chapter 4

POINCARÉ'S MAPS

Time, Reason, Nation

A FTER THE 1884 World Time Conference set the prime meridian in Greenwich, resistance in France hardened. Janssen returned to Paris, still fuming from the rout. Appearing before the French Academy of Sciences on 9 March 1885, he recounted, blow by blow, the battles of the previous year, starting with the political: the Americans had loaded the meeting with small states that were allied to the U.S. Happily enough, there were some victories, he assured his colleagues, reprinting a long speech by the French delegation. One after the other, Janssen recalled, the Americans and British had taken the floor to fight the French, each Anglophone in his special domain of competence. "It is perhaps permitted to say, despite the authority, the talent, and the number of scientists who combated the principle of the neutrality of the meridian, that principle resisted these shocks without being disturbed and without any scientific breach. The meridian proposed by France remains still the impartial, scientific, and definitive solution to the question. We believe that there was honor to our country to have defended that cause."[1] On the Continent, Janssen was not alone in his discontent. The Abbe Tondine de Quarenghi campaigned in 1889–90 in the name of the Bologna Academy of Science for the prime meridian to be moved to Jerusalem, the Universal City "par excellence," center of the three continents of the ancient world and the common sanctuary of three world religions.[2]

Unlike France, Germany was not troubled by the Greenwich prime meridian. The Germans were preoccupied with their long history of the quasi-autonomous states that had left the country struggling with a hodgepodge of mechanical and electrical time systems. It was this time *dis*unity that brought the aging General Fieldmarshal Count Helmuth Carl Bernhard von Moltke to speak, on 16 March 1891, to the Imperial German Parliament. Railroads had been key to von Moltke's celebrated triumph against France. For almost a half-century he had impressed upon his countrymen the vital role of trains in the rapid deployment of military assets. Already in 1843 he had insisted, "Every new development of railways is a military advantage; and for the national defence a few million on the completion of our railways is far more profitably employed than on our new fortresses." Von Moltke drove these plans to completion, grounding his military strategy in the power of new railway lines. By the fall of 1867 he claimed that with the south German states he could have 360,000 men massed in three weeks and 430,000 in four.[3]

Such planning paid. Not only the Germans but also their French adversaries recognized after the Franco-Prussian war of 1870–71 that von Moltke's dextrous use of precision-synchronized trains had destroyed the Second Empire, fundamentally changing the balance of European power. For the twenty years following his triumph over France, von Moltke's (and later Schlieffen's) Great General Staff oversaw a massive expansion of the military as it grew into the force of a unified Reich. Patiently, obsessively, the generals ran an endless series of technical war games as they practiced the choreographed marshaling of 3 million soldiers using a hundred thousand train cars. In 1889 the military pleaded with the Reichstag to adopt standard time to simplify their train scheduling. The politicians refused.[4]

Von Moltke of March 1891 was an unequaled hero in Prussia. When he entered a public place, men stood in silence until he took his seat. So when the general appeared before a plenary session of the Parliament on the subject of time and railroads, it was a major event.[5] In his scratchy voice (he died just over a month later), von Moltke intoned:

> That unity of time (*Einheitszeit*) is indispensable for the satisfactory operating of railways is universally recognized, and is not disputed. But, meine Herren, we have in Germany five different units of time. In north Germany, including Saxony, we reckon by Berlin time; in Bavaria, by that of Munich; in Wurtemburg, by that of Stuttgart; in Baden, by that of Carlsruhe, and on the Rhine Palatinate by that of Ludwigshafen. We have thus in Germany five zones, with all the drawbacks and disadvantages which result. These we have in our own fatherland, besides those we dread to meet at the French and Russian boundaries. This is, I may say, a ruin which has remained standing out of the once splintered condition of Germany, but which, since we have become an empire, it is proper should be done away with.

From the audience rang out: "sehr wahr" (very true). Von Moltke went on to say that while the current piecemeal ruin of time might only be an inconvenience for the traveler, it was an "actual difficulty of vital importance" for the railway business and, even worse, for the military. What, he asked, would happen in case of troop mobilization? There had to be a standard, one that would fall along the fifteenth meridian (about fifty miles east of the Brandenburg Gate), that would be the reference point; local times within Germany would differ but would require an offset by a mere half-hour or so on either extreme of the empire. "*Meine Herren*, unity of time merely for the railway does not set aside all the disadvantages which I have briefly mentioned; that will only be possible when we reach a unity of time reckoning for the whole of Germany, that is to say, when all local time is swept away."[6] Empire demanded it.

Von Moltke conceded that the public might dissent. But after some "careful consideration," scientific men of the observatories would set things right, and lend "their authority against this spirit of opposition." "Meine Herren, science desires much more than we do. She is not content with a German unity of time, or with that of middle Europe, but she is desirous of obtaining a world time, based upon the meridian of Greenwich, and certainly with full right from her standpoint, and with the end she has in view." Farms and factory workers could shift their clock starting times as they wished. If a manufacturer wanted his workers to start at the crack of dawn, then let him open the gates at 6:29 in March. Let the farmers follow the sun, let the schools and courts make due with their always loose schedules. Von Moltke wanted a nationally coordinated clock based on Greenwich. What mattered to the General Staff was that the railroads and armies should answer to a coordinated single time, one linked to the emerging electric worldmap. Much of Europe followed.[7]

But not all Europeans. Perhaps the best known action against Greenwich is also one of the murkiest. On Thursday 15 February 1894, a young French anarchist, Martial Bourdin, bought a ticket from Westminster Bridge to Greenwich. According to one of two observatory assistants, when chatting in the lower computing room, the pair "were suddenly startled by a loud explosion, the detonation of which was sharp and clear. . . . I immediately remarked to Mr. Hollis, 'That is dynamite! Spot the time'." Trained to observe by the clock, they duly recorded the detonation at 4:51. When a policeman arrived at the detonation scene in the park below the observatory, he found Bourdin dying. The anarchist had lost his hand and received a massive blow of explosive and bomb fragments. For years, doubt lingered about Bourdin's motives; anarchists suspected a police setup; others saw in it one more in the long series of French anarchist strikes, including one on the Chamber of Deputies in Paris (December 1893) and another in a Paris café just three days before Bourdin's demise. Joseph Conrad's version of the events in his 1907 work *The Secret Agent* remains the canvas on which these events have been seen: a dark sketch of dupes, manipulators, and careerists from which no one emerges unsullied. In Conrad's world the conniving First Secretary of a Foreign Power insisted on an attack that would frighten the class enemies beyond murder: "The demonstration must be against learning—science. The attack must have all the shocking senselessness of gratuitous blasphemy." It must strike at the mysterious scientific heart of material prosperity. "'Yes,' he continued with a contemptuous smile. 'The blowing up of the first meridian is bound to raise a howl of execration.'"[8]

Without a doubt the first meridian stood as a powerful if highly contested symbol. But even in France, where Janssen and others recoiled at Britain's arrogation of world power, there were those who were entirely supportive of setting French time to the master clock of the great Christopher Wren observatory.

Charles Lallemand, a member of the French Bureau of Longitude and an ally of Poincaré, made his support of Greenwich time crystal clear. Of course universal time (a single time for the whole world) would be an unmitigated disaster: the Japanese man in the street would surely refuse to live and work by the time it happened to be in Greenwich at that moment.[9] Lallemand insisted that time reform would have remained mired in chaos if the North Americans, "with their admirable practical sense of *Business'men*, hadn't imagined an ingenious compromise, uniting, or approximately so, all the advantages of the universal hour with that of local time": time zones.[10]

Writing in 1897, Lallemand insisted that it was an unparalleled victory in the domain of human reform that only ten years had sufficed for a simple, practical system of time zones to have conquered almost the totality of the civilized world. All of Europe now adhered to the system with the exception of France, Spain, and Portugal. What was needed for the French to join in this reform was so little: a delay of a mere 9 minutes and 21 seconds would set things right. Not only was the present system foolishly complicated, it meant, as

Lallemand sadly noted, that on the other side of the earth, there was an equatorial zone spanning 250 miles between the antimeridian lines of Paris and Greenwich in which the very date was ambiguous. Standing (or floating) in that purgatorial time zone, you could class yourself as witnessing 31 December 1899 or 1 January 1900, depending on which set of maps lay before you.[11]

For Lallemand, such ambiguity was intolerable. Objections flew fast and furious in this campaign of articles and broadsheets. Some claimed that the zones weren't "neutral," following the line taken at the contentious Washington meeting. False, Lallemand declared: Not only did the new zone system follow the standard, neutral twenty-four-hour clock, but the "vertiginous speed" of acceptance in both the new and old worlds had showed just how neutral it was. It is quite true, he conceded, that one meridial line ran through Greenwich. But that reference point was already familiar to nine-tenths of the world's sailors. Can one really say that exchanging a longitude of zero for 9 minutes and 21 seconds would cause us to lose our French originality and scientific personality? Nonsense, he fulminated. Paris had long taken the position "20 degrees east," using a first meridian on the island of Ferro. Even the much-vaunted neutrality of the revolutionary Convention that established the meter is not exact: what began in revolutionary France as a neutral measure (1 meter = 1/10,000,000 of a quarter of the circumference of the globe) rapidly deviated from that ideal as foreign nations copied a reference bar set in Paris. How, then, he demanded, could France possibly consider it a humiliation to modify its prime meridian? Some said it would render obsolete all French maps. No, Lallemand riposted, we could even overstrike the new longitude lines in a different color. He concluded that he and his countrymen would not be sacrificing their national meridian for the British. They would merely be offsetting their clocks by 9 minutes and 21 seconds for the benefit of telegraphy, navigation, and train travel. All "men of progress," he concluded, should back the reformation of time.[12]

Decimalizing Time

As Lallemand's "defection" suggested, decisions taken at the Washington Conference of 1884 reverberated through the French Bureau of Longitude. In addition to their fateful decision about Greenwich, the Washington delegates had also "expressed the hope" that the definition of the astronomical and the nautical days might be joined, so both would start at midnight. Astronomers had long started their official day at noon, profiting from the absence of anything of observational interest during the day to keep their precious nights unbroken by a calendar change. Meanwhile the rest of the world utilized nighttime quiet to change its day at midnight, thereby keeping daylight within a single calendar day. Boosted by concern emanating from the Canadian Institute and the Astronomical Society of Toronto, the Minister of Public Instruction asked the French Bureau in 1894 for its opinion.

Poincaré led the charge and began, characteristically, by weighing inconveniences. "It is evidently inconvenient [*incommode*] for an astronomer to change the dates in his notebook in the middle of a night of observations." He could forget to make the switch, leaving records vastly complicated. But this same "incovenience" now exists for the sailor making solar observations at sea. Indeed at every moment the sailor currently faces precisely the same problems as the astronomer, and worse. Where the astronomer works without distraction, the mariner has a million worries; where the astronomer can always give the observations a fresh look later, the mariner could be dashed against shoals by the slightest error. So let the astronomers write "night of 11–12" in their notebooks. What about the discontinuity in time on the day the reform was instituted? Better to straighten out that "inconvenience" now than later, Poincaré responded.

If there was a serious objection, it was this: any such reform would necessarily scrap the many compilations of astronomical phenomena, such as those inscribed in the British and American astronomical almanacs, the French *Connaissance des Temps,* or the German *Berliner Jahrbuch.* Were any of these volumes to alter their time conventions, converting among them would present a vastly greater problem than the current confusion. Only an international accord could implement this laudable effort to simplify time. Until international accords prevailed, the Bureau's members concluded that the French should wait, though all agreed that a twenty-four-hour clock would be an improvement.[13] But soon the French put even the twenty-four hours in a day under scrutiny, following up on the Washington Conference concession that left "hope" for time's decimalization. In February of 1897 the president of the Bureau of Longitude established a commission with Henri Poincaré as its secretary to determine whether France should scrap the old twenty-four-hour day and 360-degree circle in favor of a truly rational system. The president bluntly asked: why had the great Convention of 1793 failed to extend the decimal system to time and the circumference of a circle, and were objections to such a novel system still valid?[14]

One commissioner serving with Poincaré, Bouquet de la Grye, a former Polytechnician and prominent hydrographic engineer, responded. The Convention had aimed to banish everything that recalled the ancient measuring customs of the ancien regime and to consecrate true French unity by deploying common measures across the whole of the country. With the meter, the Revolution succeeded. But the Convention's time reform of months and weeks had been a disastrous flop, as had Laplace's valiant but doomed attempt to decimalize the hour. De la Grye reminded his colleagues that few decimal clocks survived, and no one outside of France took any interest in them. From this debacle, de la Grye found clear guidance for the present: "The metric system succeeded because it was the simplest and it put an end to a veritable incoherence in local measures; the decimalization of time and circumference failed because the whole world employed the same measures and the proposals sinned precisely by their lack of unity."[15] Convenience ruled. Where reforms simplified lives, the public went along; when reforms did not help ordinary people, the schemes fluttered softly into oblivion.

Poincaré duly recorded the debates that followed. President Loewy insisted that if the Revolution's metric time had failed, it was because French astronomers could not find partners for the reform in other European states. General de la Noë noted that the geographical service had indeed adopted decimal angles, as had the geodesic services in Belgium. Cornu ascribed one difference between the late-eighteenth-century and the late-nineteenth-century present to the loss of habituation to the duodecimal system. The most curious feature of all was that the contemporary British engineers were so trained in their primitive measures of "inches" and "feet" that they simply refused to understand the advantages of the decimal system. Less despairing of his cross-channel colleagues, the director general of the Paris-Lyon-Mediterranean Railroad evinced some sympathy for the Anglophones; he reported that many British engineers suffered under their present system and would love to break out of their island's archaic conventions. Impressed by the historical French drive toward the revolutionary ten, stirred by their historic role in rationalizing the world, the commission voted to decimalize time.

But that vote left much undecided. The railroad representative assured his scientific colleagues that any attempt to alter the twenty-four-hour day was doomed to failure. De la Grye held out for a unification of time and geometry by dividing the circle into 240 parts. Loewy admitted that he had dreamed of a full-bore decimal world, but the burden of past maps and measures led him, mournfully, to conclude that the obstacles to such a brave new world were not passing difficulties. He endorsed de la Grye's compromise; a circle divided into 240 parts would put the globe into synch with the clock as each hour of the earth's rotation would correspond to a turning of the world by an elegant 10 degrees. Poincaré backed Loewy, adding:

> Were we in the presence of a tabula rasa, the best system would be one that divided the circle into 400 parts [100 grads per quarter circle, or 100 kilometers per grad]. . . . But we cannot break completely with the past, because not only must we take account of public repugnance, but scientists themselves have a tradition to which they remain tied.[16]

Keep the twenty-four-hour day, Poincaré said, and divide the circle in 240 parts. Commander Guyou rejected such compromises, insisting on division into 400 parts. The navigational demands at sea and the calculation of tides brought the mariner into a never-ending series of painful and difficult calculations. Why not provide an easy-to-use decimal system for the scientist or navigator while leaving the public to count time according to its old habits? Trainmen, Guyou added, were already habituated to a timekeeping system that was off by five minutes from city time and were equally at ease with a twenty-four-hour clock. Why not a two-time system? Such an amalgam was intolerable, Poincaré replied. Civil time was clearly linked to longitude. Any mixed system necessitated a conversion factor that was truly convenient for any user.[17]

Convenience, convention, continuity with the past. These terms arise again and again in Poincaré's abstract philosophy. Yet here they are written in the less-than-etherial concerns of real-world engineers, seagoing ships' captains, imperious railroad magnates, and calculation-intensive astronomers. Monsieur Noblemaire, director of the Paris-Lyon-Mediterranean Railroad, argued as follows to reinforce his intervention: Suppose you leave your train station at 8:45

A.M. and arrive at 3:24 P.M. How long is your trip? One has to think. But by expressing travel problems in decimals, the confusions of A.M. and P.M. vanish. All that remains is subtraction:

    start: 15.40h
    end: 8.75h
    duration: 6.65h

Here was convenience that mattered to a society on the move.[18] Mariners, like the Commandant of the Naval School, saw only benefits to the transition to decimal time and longitude. Maps would be easy to alter, and physicists shouldn't have a problem. After all, according to the good captain, physicists had adapted easily enough to centimeters, grams, and seconds.[19]

Decimalizing time, easy? That would have been news to the physicists. When the president of the French Physical Society, Henri Becquerel, got his hands on the proposal in early April 1897, he was not amused. Even leaving aside the expense of changing all clocks, marine chronometers, pendula, and watches, the physicists saw dire consequences for the electrical industries and those that flowed from them. For it had only been in 1881 that the centimeter-gram-second system (CGS) had been adopted internationally and only in April 1896 that the president of the Republic had decreed that the rational CGS system of measure should be used in all matters of State. Needless to say, one of the cornerstones of that newly rational system, the sexigesimal second (1/3,600 of an hour), now lay in the crosshairs of this decimal revival. Not only would the proposed decimal second (1/10,000 of an hour) utterly alter the mechanical and electrical units of current, work, and the like, but all the practical units (amperes, volts, ohms, and watts) that are defined in terms of them would have to change as well. "What an upset in scientific practice and the entire mechanical and electrical industry!" All the instruments would have to be altered. "What a huge expenditure without any profit, neither for science, nor for industry!" If one weighed the advantages against the disadvantages, the physicists' balance tipped decidedly toward the status quo: keep the old second. For the physicists, this case was closed before it opened.[20]

The physicists' plaintive cries did not move Poincaré. Or rather, in the face of fierce protests from the public, the navigators, and the scientists, Poincaré refused to engage the debate at all, instead proposing to resolve the problem with the technical legerdemain of a reforming Polytechnician. On 7 April 1897, Poincaré brought the commission a table he had prepared that displayed each proposed system, the kinds of multiplications required (ignoring factors of ten) to express angles, to convert time to angles for expressing longitude (degrees per hour of the earth's rotation), and to convert angles from the old 360-degree system to the proposed decimal ones. For example, to express an angle of a circle and a half (1.5 rotations) in the 100-division system, there's no conversion factor at all: once the factors of ten are taken into account, you just have 150 units. No mental arithmetic, the multiplier is just 1. For the 400-division system, 1.5 circles would be 600 units; to get from 1.5 to the 6 in that 600 (the additional multiplication by 100 requiring no brainpower), you need to multiply by 4. The second column tells how to switch from parts of the circle into time: if the whole circle is 100 units, you need to multiply by 24 to get hours—100 units of rotation equals a full twenty-four-hour day. If the circle is 400 units, you need to multiply by 6 to get hours. Finally, to switch back into the old 360-degree circle, you would (for the 400-unit circle) just multiply by the factor 9, whereas a 100-unit circle required multiplication by 36.

Good technocrat that he was, Poincaré had scanned the table to find the simplest conversion factors. Only the 400-unit system required no double-digit multiplication. So there it was. Poincaré had, objectively, the least "inconvenient" solution, one he happily defended to shrieks of protest against the proposed scrapping of every angle-measurement in the world.[21]

Divide circle into this many parts	Factor for angles greater than circle	Factor to convert arcs into time	Factor to convert to 360 degrees
100	1	24	36
200	2	12	18
400	**4**	**6**	**9**
240	24	1	15
360	36	15	1

Poincaré's table presenting the case for dividing the circle into 400 parts. Source: Henri Poincaré, "Rapport sur les résolutions de la commission chargée de l'étude des projets de décimalisation du temps et de la circonférence" [7 April 1897], Archives of the Paris Observatory.

Here was an engineering-dictated armistice in a war pitting all against all on social, economic, and cultural grounds. When the dust settled, the hostile factions had compromised even further, keeping the twenty-four-hour clock and decimalizing the hour into 100 minutes with each minute split into 100 seconds.[22] These half-measures left Captain Guyou cold. Sea captains were habituated to reading complex tables, as he drily noted; it mattered little to him whether the table consulted had simple or complex formulae for conversion. Ultimately, the committee splintered into almost as many opinions as there were members. One camp lobbied for a division of the circle into 400 parts, another wanted 240 parts, and a third (physicists, sea captains, and telegraphers) preferred the tra-ditional 360 divisions with decimal subslices. Astronomer Faye bucked all these proposals, demanding a clean split of the round into 100 parts.

Amidst this deci-strife, Poincaré and his allies tried to remain above the fray, adjudicating the competing systems with equanimity. Scribe and contributor Poincaré dutifully recorded his own opinion, one entirely consistent with his views about unifying civil and astronomical time: all these "systems are acceptable and one must choose that one which would have the greatest chance of success before an international congress." With the president on his side, Poincaré's forces won in a voice vote: the unity of angle would be the "grad," that is to say 1/400 of the circumference. That decision, while reinforced in a later meeting, did not quiet the debate. Some of the commissioners themselves dissented: the chief engineer of the French hydrographic service filed a report protesting the tremendous burden that would be caused by trying to reprint the 3,000 charts (not to speak of the instructions, tables, and yearbooks), issued by the service. The entire instrumentarium of navigators, moreover, would be rendered obsolete overnight. Sailors might as well throw their chronometers, pendula, watches, theodolites, and sextants into the briny deep. Another protester contended that the decimal commission was merely managing contradictory interests in a way that would disappoint the public with a badly tailored compromise. Cornu argued that there was one rational system and only one: the proposed reform was neither an adventurous, rational charge into the future nor a safe retreat to the status quo. For Loewy, the radical decimalizers had missed the point: the new system cut out much of the irrational and complicated bits of a system that, in its historical bricolage, made no sense. For the moment one could do no better than settle. Loewy and Poincaré had the votes to back their liberal compromise of a partial reform: 12 for, 3 against.[23]

Voting silenced no one in the debate over the measurement of time, which tumbled, later in 1897, into the pages of Cornu and Poincaré's journal, *L'Eclairage Electrique*. Clearly uncomfortable with the commission's compromise, Cornu reported that "an unstable and divided majority" had formulated a solution that would have great difficulty in finding universal acceptance. Even before the committee had finished its work, he said, the French Navy and the army's geographical service both had objected, since the former would gain little simplification in their usual calculations, while the latter would actually be stepping backward. According to Cornu, decimalizing time was harder than decimalizing space. Length reform had met a triple condition: it held clear benefits for a majority, it offered no enormous inconvenience for those not directly concerned, and it fit well with general public enthusiasm for a unity of length. Everyone was glad to escape from the trading confusion that had prevailed at national, provincial, and commune borders. He expected no such celebration of this flawed time reform.[24] Cornu insisted that it was the day, the *natural* unit of time, that should be decimalized—not the wholly artificial hour. If the day were the base, then a hundredth of the day would be just about a quarter of an hour, and a hundred-thousandth of a day would equal 0.86 old-style seconds. That would be a gratifying unit of time because it corresponded so closely to the typical adult heartbeat, our "natural" small temporal unit.

But "interests trump logic," Cornu dourly noted, and no interests backed the one logical reform that would bring order to time. Chaos did *not* currently reign in the world of time, as it had in spatial measurements before the reform of the meter. In a sense, time was already unified among countries, as length had *not* been before the meter. Into this unreceptive atmosphere, Cornu observed, the time commission was now dropping a hopelessly confused compromise that sanctified the all-too-artificial number of twenty-four hours in the day, and then decimalizing this useless hour. Decimal fractions of an hour were unnatural, Cornu insisted: a hundredth, thousandth, and ten-thousandth part of an hour were 36, 3.6, and 0.36 seconds, respectively. This was no good; an astronomical clock could not beat at intervals corresponding to these units of time: 3.6 seconds was too long for a pendulum and 0.36 too short. Having spent years building and maintaining his own astronomical clock with a massive pendulum, Cornu charged ahead: the whole human organism made the second special. Not only did our pulse surge every second or so, but our capacity to react, by sight or sound, took about a tenth of a second, adding to the value of a time unit that was approximately a heartbeat long. According to Cornu, the body, the existing clocks, and the sun all militated against decimalizing the hour. Cornu despaired, moreover, at the commission's other compromises. If the hour was to be maintained (which he opposed), then the logical division of the globe ought to be in 240 parts, making each hour of rotation carry the earth through the even division of ten parts. Yet here, too, the commission had misplayed its hand, dividing the circumference of the world into 400 parts. Since 400 parts did not divide evenly into the twenty-four hours, even the geographers would gain little convenience as they reckoned longitude. As far as Cornu was concerned, this reform had failed to decimalize the only truly natural period of time by refusing to seize the day as the basic temporal unit in our lives: "men of science should . . . prepare the future not by compromises, but by the progressive adoption of [the decimalized day and circle] in the domain in which they are masters of convention."[25]

Attacked in public by his friend and mentor over the compromise solution that he had crafted, Poincaré too now had to intervene in print, not least because the legitimacy of his commission had been thrown into question. Like Cornu, he acknowledged the competing interests. Unlike Cornu, however, Poincaré saw such discord as the mandate for a rough and ready settlement. As always he aimed for an engineer's progressive middle course, avoiding both revolution and reaction. The essential point, in his eyes, was to drum out of existence such monstrosities as 8h25m40s or 25°17' 14". As far as he was concerned, the important point was the adoption of *any* decimal system.

As Poincaré knew full well, the physicists (that is, the electricians) had fled the room. Poincaré now tried, gently, to usher them back. Surely they had exaggerated the inconvenience of the decimal system. If they'd only read the report, they would come over to his side (or so Poincaré thought). So he began to cite from it. After so many years of struggle to establish an acceptable system of electrical units, it was understandable that the electricians could not easily abandon it now. But, he urged them, be reasonable. For all practical purposes, by any measurement one simply is comparing one thing to another: one time to another time, one resistor to another resistor. In any such industrial operation there is no need to go back to the underlying definition of a unit. The storekeeper measures cloth with a meter and has absolutely no need to remember that the meter is the forty-millionth part of the terrestrial meridian. Only the so-called absolute units would be damaged by the reform. (An absolute unit of electrical current, for example, was defined not in relation to any particular materials, but by identifying an ampere as just that amount of current which when passing through two parallel, infinitely thin bars a meter apart would repel each other with a certain force.) In effect, Poincaré said: Let the English worry about the natural theology of absolutes; it was convenience, not divine sanction that mattered.

Indeed, Poincaré insisted that the object of the physicists' complaint was inconsequential. True, the old sexagesimal clocks with their sixty seconds and sixty minutes could only awkwardly be compared to the new observatory clocks that would display centihours (hundredths of hours, equal to thirty-six seconds). So what? Chronometers need only indicate intervals of time, indeed very small intervals. There is simply no need to reset them to a specific time of day. But push the thought experiment to the limit. Imagine, Poincaré continued, that the astronomers adopted the decimal system, followed by such a spectacular diffusion among the general public that there was not a clock based on seconds still to be found anywhere. How much bother might those rare physicists be subject to who wanted to determine the absolute value of electrical resistance (the ohm)? They would have to multiply by 36. "And for them to avoid this operation we are daily to impose tedious calculations on thousands of mariners along with millions of schoolchildren and former schoolchildren?" Which do we do more often—determine the absolute value of the electric resistance, fix our position at sea, or add two angles, or two times? In sum, Poincaré accused the physicists of blocking progress for the astronomers and the public just because they, the physicists, would not profit. As far as Poincaré was concerned, some progress was better than none. If units differed in astronomical treatises and electrical books, that was a small price to pay to rid the world of the absurdity of having three units in a single number like 8h,14m,25s.[26]

Despite the prodigious amount of work they put into their campaign, Poincaré's time commission stalled. Finding open foreign hostility to the idea of such time reform, the Ministry of Foreign Affairs let the Bureau of Longitude know in July 1900 that State was not ready to back the effort. After a hundred years, the revolutionary struggle to rationalize time expired.[27]

Although they lost the attempt to decimalize time, many participants in Poincaré's commission argued passionately over time zones and time distribution. Sarrauton, for example (never one to avoid polemics), turned his guns toward the zone system. He took one of his scathing reviews (on the proposed time zones) and sent it personally, on 25 April 1899, to Loewy at the Bureau of Longitude. It began, as so many time-pleas did, with an homage to the rail and cable: "The surface of the globe is criss-crossed by trains riding on rails, by fast boats full of voyagers and merchandise, and on the aerial and submarine telegraphic lines, news circulates with the speed of light. . . . The surface of the planet has been, in a certain sense contracted." Time zones of synchronized clocks were the answer, but certainly not *these English* time zones.

Sarrauton demanded that the wedge-shaped time coordinated divisions of the earth be liberated from the grasping claws of the British Empire. His ire had been roused by a proposed "Boudenoot" law that would have delayed Paris time by 9 minutes and 21 seconds: "This is immediately the time of Greenwich; the English meridian, soon, 'France towed by England' and the collapse of the metric system!" Happily, as far as Sarrauton was concerned, there was an alternative—the law of Gouzy and Delaune—that would do the right thing by zones, decimalization, and the long-lost prime meridian: it would divide the hour into 100 minutes, the minute into 100 seconds, set civil time into twenty-four time zones, and count longitude from the Bering straits beginning on 1 January 1900. "This is the achievement of the system of decimal units, France realizing one of the most important reforms of modern times, and in scientific matters, the preponderant French influence in the world. We have arrived at a crossroads and these two projects of law mark the two roads that open before us. We must choose."[28] Whatever its advantages for rational France, Gouzy and Delaune's proposal never passed. France adopted the Greenwich meridian on 9 March 1911.

In these ferocious debates over zones, decimalization, and the prime meridian, time conventions crossed arenas that we tend to think of as far from each other. Laws, maps, science, industry, daily life, and the legacy of the French Revolution all collided, drawing in leading figures from the French technical, intellectual, and scientific establishments. At the Bureau of Longitude, Poincaré's philosophical hope for "conventions" and "convenience" ran smack into the everyday realities of navigators, electricians, astronomers, and railroaders. Just before Poincaré published "The Measure of Time" in 1898, physical, conventional, and coordinated time had converged in whirling reformist debate that was thoroughly abstract and yet altogether concrete.

Of Time and Maps

If one of the Bureau of Longitude's projects in 1897 was the establishment of decimalized time, even more pressing was the orchestration of one of the most difficult time-synchronized mapping projects in the Bureau's illustrious history. Already in 1885, the Ministry of the Navy had assigned the Bureau the task of determining the exact positions of Dakar and Saint-Louis in "our colony" of Senegal.[29] Years in the making, the Senegal report only appeared among the Bureau's publications in 1897—coming into Poincaré's hands just before he wrote "The Measure of Time" and took on the presidency of the Bureau.

Mapping "our colony" was no easy task. Governor Louis Faidherbe had violently annexed the Wolof and Cayor regions of Senegal by 1865, enabling the colonizers, at least intermittently, to clear the road from Saint-Louis to the Cape Verde peninsula. By 1885, the colonial government had a railway link under construction joining Saint-Louis and Dakar.

But laying iron tracks did not stop fierce anticolonial resistance. French forces, still battling in the east and south, never completely suppressed the rebellion against them. Insurgencies were still erupting on the eve of World War I. In the heat of these colonial battles, the Bureau's men were there to fix the position of Senegal's two main cities, in part to extend mapping into the interior of the colony alongside conquering troops. But the French colonial project had even grander ambitions. By providing exact coordinates at Dakar, the French authorities intended to extend their cable system from that port down the length of the west coast of Africa to the Cape of Good Hope. Longitude, train tracks, telegraphy, and time-synchronization reinforced each other. Each showed a different facet of a new global grid.

By the time the Dakar–Saint-Louis expedition set off from Bordeaux, the Bureau had already extended its telegraphic reach to Cadiz's San Fernando Observatory (about fifty miles northwest of Gibraltar). From there, the French relied on British Cable's link joining the Cadiz and Tenerife observatories (Tenerife Observatory had just recently been cabled to Senegal). One hour each evening,

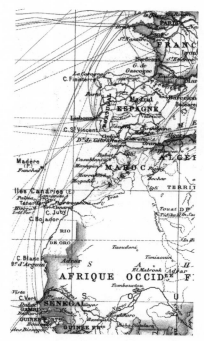

Figure 4.1 Cabling Simultaneity: Paris, Cadiz, Tenerife, and Dakar.
Undersea cables proliferated in the last part of the nineteenth century, and
longitude finders instantly seized the opportunity of using each new link to
wire time across the oceans. The French Bureau of Longitude made use of
British and Spanish submarine cables to get their time signal from the obser-
vatory at Cadiz, Spain, over Tenerife and thence to Dakar. SOURCE: VIVIEN DE
SAINT-MARTIN, *ATLAS UNIVERSEL* (1877).

the astronomers controlled the cable. For France, as for all of Continental Europe, renting cable time from the British had entered
into the order of things. Controlling the vast majority of the world's
submarine cables, the British passed messages between France and
its colonies everywhere but North Africa. Alone the Tenerife-
Senegal, West African, Saigon-Haiphong, and Obock-Perim cables
cost France almost 2.5 million francs per year, paid, gallingly

enough, to their imperial competition, the British. Such dependence burnt slow and deep in the French establishment; the military, commercial, and journalistic sectors all chafed during the 1880s and 1890s under British communications dominance. But the Chamber of Deputies blocked French cable proposals one after another: One casualty was an 1886 link from Reunion to Madagascar, Djibouti, and Tunis; another, in 1887, from the French West Indies to New York, and a third from Brest to Haiti, in 1892–93.[30]

Just as they relied on the British cable-laying firm, the French authorities needed the Spanish. Cecilion Pujazon, astronomer at San Fernando, offered his observatory as a relay point, linking the Paris-launched signal to Senegal through Tenerife. Taking the steamship *Orenoque* of the Compagnie Transatlantique, the astronomers arrived, worse for the wear, in Dakar on 15 March 1895. Governor Seignac-Lesseps immediately put his troops at their disposal. The artillery captain commanding the Dakar garrison offered laborers and indigenous masons, and housed the scientists in the military's mess halls—there being (according to the querulous chief astronomer) no decent hotel to be found anywhere. With horror the chief astronomer noted that passengers were sleeping in straw bedding like natives.

Military aid to the map troops did not end with food and lodging. The Bureau of Longitude astronomers took up their posts in the fort's blockhouse, looming on the point of Cape Verde to protect the coal park and anchorage. Thick concrete walls designed to protect artillerists against attack allowed the astronomers to install their timekeeping pendulum in the bunker's sheltered room. This kept the temperature of their all-important clock under control, an absolute necessity in Dakar's ferocious heat. Establishing their meridian telescope outside the blockhouse, astronomers mounted the collimator on the parapet of the gun battery. Five days later, on 20 March, the team steamed for Saint-Louis. Welcomed by the governor's aide-de-camp at "the political center of our colony," they hammered into place a Saint-Louis station.[31]

A gubernatorial edict prevented people from entering or leaving their operations building during the observations (footsteps disturbed the pendulum). Those few natives allowed to pass by did so without carts, in bare feet, on sandy soil. Observations ran from 26 March until 11 April 1895; the team launched their first signals to Dakar on 29 April and to San Fernando on 2 May.

Not everything went well. The river flowed north-south, making the Saint-Louis meridian sighting extremely difficult. It was good to

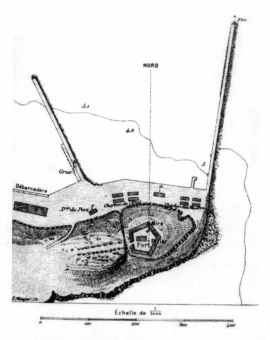

Figure 4.2 Measuring Dakar. The overlap of military structures and geographical work was extensive—for the French, the British, and the Americans. At Dakar, the French longitude expedition used every aspect of the fort overlooking the vital coal yards to establish the fundamental longitude measurement of the colony. SOURCE: ANNALES DU BUREAU DES LONGITUDES, VOL. 5 (1897).

be on the riverbank to avoid obstacles, but they couldn't very well plant themselves in the middle of the river to sight due north toward the polestar. "Marauders" began stealing anything metal left in the open, including, almost daily, the spike used for collimating the shooting of the stars.

> The 260 kilometers of wire that link Dakar to Saint-Louis were established in the bush-covered plains of Cayor, with primitive means and in the midst of a population that was hostile if not in open war with us. The wire was so often cut, the poles so often knocked down and then re-established more or less well, that one does not find the normal [physical] conditions of an ordinary line. Let me add that at sunset the dew was so strong that the poles dripped, and where 5 insulators ordinarily would suffice [back in France], even 70 failed to give a trace of electrical wave.[32]

Even once the signal made it along the train tracks through Senegal, the signals from Santa Cruz (Tenerife) to San Fernando failed because, during the crucial nights of observation, there were active and official transmissions on the Spanish land lines following the election of deputies. Then came corrections. There were the usual corrections for temperature variations; there were corrections through "personal equations," the characteristic psychophysiological lags of each observer as he signaled a star crossing the meridian. There were corrections for the behavior of the delicate mirror that registered the signal's arrival. In the end, after doubling the weight of one set of observations and compensating for various other instrumental and human errors, the team settled on a San Fernando–Saint-Louis longitudinal difference of 41 minutes, 12.207 seconds.[33] Not wanting to miss the opportunity of a rare steamboat leaving Dakar for Tenerife, the astronomers corralled some homesick sailors to quickly pack their instruments for the voyage back to Paris. Their report to the Bureau appeared in print in 1897.

By the time the Senegal report saw light of day, French-British relations over their colonies were deteriorating, and telegraphic cables ran through the center of many disputes. British cable companies refused to employ foreigners and were clearly reading French messages between Paris and Dakar or Saigon even before the French authorities had them in hand. At the 1885 Cable Convention, Britain insisted on the rights of warring nations to tear up their opponents' lines — clearly an advantage to the country that owned twenty-four of the thirty extant cable ships. In 1898, tensions increased to the brink of war. Racing to control a part of the Nile in Sudan, Captain Marchand's French squadron was ready to stake a claim; but while the British troops remained in continuous contact with London, the French cable inexplicably went silent. Only when the French Governor loaded the cannon in Dakar (where the French Bureau of Longitude men had just been) did the cable, magically, come alive.[34]

Back in Paris, the longitude battle between the French and British capitols would not end and could not be contained in the domain of the politicians. Even as the Bureau of Longitude continued to vex itself about whether French clocks should be set by Greenwich time, the astronomers went clock-to-clock over the most basic of questions: where was Paris in relation to London, or rather, how far east did the Paris Observatory stand in relation to the Royal Observatory? Back in July 1825, the two national rivals had tried to settle the problem by launching rockets from the English Channel, using their explosions to synchronize clocks. Le Verrier and Sir George Airy then tried again by telegraph in 1854, settling on a difference of 9 minutes and 20.51 seconds. Unfortunately that consensus collapsed in 1872 with the American Coast Survey telegraph campaign that accompanied the laying of the transatlantic cable. It determined that Paris was nearly a half-second farther away from London — 9 minutes and 20.97 seconds. Half a second was far too large an error for anyone to tolerate in the era of hundredths, if not thousandths, of a second, so the Astronomer Royal, along with General Perrier (director of the Geographical Service of the French Army) and Admiral Mouchez (director of the Paris Observatory) joined forces in 1888 to settle the question (so they hoped) for good.

Appointing two observers from each country, the astronomers planned to conduct measurements shoulder to shoulder. A Frenchman and an Englishman would literally stand next to each other, with one pair in England and the other in France. They shared a telegraph line, they traveled together back and forth across *La Manche*, and they coordinated their procedures down to fine corrections of personal errors. They even shared a single battery-run electric light to avoid needless heat that might distort their instruments. At Montsouris, the French and English instruments stood on piers twenty feet apart. Yet the results caused dismay:

English longitudinal difference from Paris to London:
9 minutes and 20.85 seconds
French longitudinal difference from Paris to London:
9 minutes and 21.06 seconds.

A fifth of a second. Still too much. So again, in 1892, the year Poincaré was elected a permanent member of the Bureau of Longitude, after years of negotiation, the teams once more wired their telegraphs into place. Again the astronomers secured their instruments on the stone piers, launched electrical waves, and painstakingly reduced data. To their immense embarrassment, the results were no more harmonious than they had been in 1888. The two greatest European observatories, both of which claimed to map the whole world, could not agree on their own positions to within a fifth of a second. Once more the French found a greater expanse between Paris and London than the British did. This telegraphic time crisis finally came to a head in 1897–98, during the Bureau of Longitude presidencies of Loewy and then Cornu. The Paris Observatory, in conjunction with the International Geodetic Conference, insisted that the two observatories stabilize the European map with a proper time exchange.[35]

Clearly, issues of conventions were not confined to the far reaches of geometry or philosophy, rather, they were ubiquitous. Conventions concerned the prime meridian, decimalized time, submarine cables, map making—even the relative location of Paris and London. Everywhere one looked from the Bureau of Longitude in 1897–98, international concords in the world of space and time seemed urgent.

In this light, Poincaré's "Measure of Time" reads as something quite different from a purely metaphorical tract. First, the calculation of Paris time from a distant site (whether London, Berlin, or Dakar) was not an abstract problem at the Paris Bureau des Longitudes in 1897: it was the most pressing issue of their map-making mission. Second, 1897 marked an enormous intensification of the long war over conventional decimalization, a debate that Poincaré engaged directly. In short, in 1897 more than ever, it was the Bureau of Longitude that was responsible for the precision simultaneity measurements that would permit the extension of an ever-more accurate map of the world relative to the marble longitude pier of Montsouris.

Indeed, when Poincaré proposes in "The Measure of Time" that simultaneity must be understood as a convention, it is crucial to attend to the literal. Synchronizing distant clocks by the observation of astronomical events was standard practice for French, German, British, and American surveyors. Transits of Venus, occultations of stars by the moon, eclipses of our moon and Jupiter's moons were all useful events to set clocks by (if only inaccurately) on distant colonial shores. Having been a member of the Bureau for four years in 1897, Poincaré knew perfectly well that clock coordination by astronomical sightings had, for many reasons, long been overtaken by the telegraph as the most accurate standard of simultaneity.

Instead—this is the crucial point—Poincaré invoked telegraphically determined longitude as the basis for establishing simultaneity between distant sites. He insisted, in the most celebrated lines of the essay, that in the synchronization of clocks one *had* to take the time of transmission into consideration. He immediately added that this small correction makes little difference for practical purposes. And he remarked that the calculation of the precise time of transit for an electrical telegraph signal was complex. From at least 1892–93, Poincaré had taught the theory of telegraphic transmission of signals and reviewed the experimental studies that measured the speed of electrical transmission in wires of iron and copper. That interest did not wane. In 1904, in a series of lectures to the Ecole Supérieure de Télégraphie, he extensively analyzed the "telegrapher's equation," comparing it to work by others and specifically referring to the physics of undersea telegraph cables.[36]

In the practices of signal transmission time lies a key point to our puzzle. At first glance, it might seem impossible that Victorian cartographers were taking account of the time of signal transmission as their electrical pulses sped simultaneity across continents and oceans. But we need to look at what they were actually doing as they struggled against the myriad errors that entered the procedure. In fact, in their error corrections, the electric mappers had long since been taking precisely the transit time into account as they synchronized clocks between Paris and the far reaches of the United States, Southeast Asia, East and West Africa. Or, for that matter, in the intransigent precision measurements between Paris and Greenwich. They did not need to wait for relativity.

Back in 1866, as we saw, the U.S. Coast Survey put a monumental effort into determining the longitude gap between Cambridge, Massachusetts, and Greenwich, England, over the first Atlantic cable. The surveyors corrected for the usual errors in clock rates and in the measurements of star positions. But they were quite aware that the instant a telegraph key was pushed in Calais, it did not register in Valentia. That gap was in part due to the observers— their reactions were not instantaneous—and in part due to the instruments' inertia. It took time, for example, for the magnet to swing the little mirror around sufficiently to cause a noticeable deflection of the light beam. These difficulties, which the authors dubbed the "personal error in noting," could largely be eliminated by measuring these delays under controlled circumstances. But there was another crucial contribution to the delay between transmission and receipt, and that was the time it took for the signal to cross the Atlantic from Nova Scotia to Ireland. Amidst the listing of all the other errors, there in plain print are the measured transmission times: 25 October 1866, 0.314 seconds; 5 November 1866, 0.280 seconds; 6 November, 0.248 seconds.[37] Or look back at the Coast Survey's myriad missions to Mexico, Central America, and South America. Or to de Bernardières's missions, or to the vast web of links among European observatories.

Everywhere the electrical surveyors were measuring the time it took the telegraphic signal to pass through the wires; *everywhere* surveyors were using that delay correction to establish distant simultaneity. Here is how they reasoned. For simplicity, let's grossly exaggerate the time the signal takes in transit, and call it 5 minutes. Suppose the eastern observing team sends its signal at 12:00 noon, eastern local time, to their western partners located 1/24 of the way around the globe—that is, exactly 1 hour westward (−1 hour). Because of the transmission time, the westerners would receive the signal at 11:05 A.M., western local time. If they forgot to correct for the 5-minute delay, this naive western mission would conclude that they were at a longitude point that was 55 minutes earlier (−55 minutes) from the eastern stations.

(East-to-west apparent difference) = (real longitude difference) + (transmission time)

(Here −55 minutes = −60 minutes + 5 minutes.) Now let's imagine what happens for a signal traveling west to east. West sends at local west noon (1:00 P.M. local time in the east). But by the time the signal from the western station arrives in the east, the eastern clock would not read 1:00 P.M., but rather 1:05. Our observers, if they were unaware of the time it took the signal to travel, would conclude that western local time was 65 minutes earlier than eastern local time (–65 minutes). In other words, the actual longitude separation would be *shorter* than our observers would believe if they forgot to take into account the signal time. This means:

(West-to-east apparent difference) = (real longitude difference) – (transmission time)

(Here: – 65 minutes = – 60 minutes – 5 minutes.)

Now, if you add the measurements (east-to-west apparent difference) and (west-to-east apparent difference), you get exactly twice the real longitude difference. The + and – transmission times simply cancel out. And, if you subtract the (west-to-east apparent difference) from the (east-to-west apparent difference) you get twice the transmission time: (apparent east-to-west – apparent west-to-east) = 2 times (transmission time). So,

Transmission time = 1/2 (east-to-west apparent difference – west-to-east apparent difference)

That simple tool for calculating transmission time formed part of the procedural mantra of every long-distance surveying team in the world: in the West Indies, in Central America, in South America, in Asia, and in Africa. Certainly the Bureau of Longitude's peripatetic astronomers had long been using it as they hauled their wooden observing shacks from place to place: between Hong Kong and Haiphong or Brest and Cambridge. Telegraphic map-makers saw, understood, and said perfectly clearly that the telegraphic signal time *had* to be taken into account to establish precise simultaneity and therefore longitude. For Poincaré *not* to have known this around 1898, we would have to assume that he ignored *every* Bureau of Longitude report written during the years he had served as a member, as well as avoided hearing any discussion of their actual procedures. We would have to suppose that when, in the "Measure of Time," he wrote "let us watch [the telegraphic longitude savants] at work and look for the rules by which they investigate simultaneity," he somehow did not understand what his own longitude crews (and every other British, American, German, and Swiss team) had been doing for the previous quarter century. This begs credulity.

Back in the late 1870s—when Lieutenant de Bernardières, Captain Le Clerc, and astronomer Loewy sought, quite literally, to reattach France to the map of Europe—one of their most important links was to be a telegraphically fixed longitude difference between Paris and Berlin. Time delay was an immediate concern. The authors noted in 1882 that their signals "were affected by small errors with different causes all of which must be taken into account carefully": errors due to the response of electromagnets, the loss of time caused by the sluggishness of the mechanical parts, the relative separation of the pen tips, and finally the "non-instantaneity of the transmission of the electric flux." To determine that transmission time, there was work to do in the observatory; the astronomer-surveyors had to be certain, for example, that the currents used were always the same. But there were also assumptions to be made, for example, that the velocity of the electric signal was the same in both directions. In agreeing to the procedures for fixing time through exchange of telegraphic signals, the language of conventions entered up front. To synchronize clocks between Berlin and Paris, protocols had to be established, agreements concluded; they even scripted in advance the greetings between stations. As in the Convention of the Meter, conventions of simultaneity demanded international standards, detailed accords on every step of the process.

Now it makes sense that we find among the headings of the French Bureau of Longitude's Paris-Berlin report, *Conventions Relative to the Exchange of Signals*.[38] When the Bureau of Longitude nailed Bordeaux to the map in 1890, in the list of other corrections, such as pendulum errors and personal equation errors, we finally come to it: "the delay S of the electrical transmission."[39] For the French team in Senegal, it was therefore fully routine to take account of the finite signal time of their electrical pulse; like so many other telegraph missions by 1897, the team simply applied the by-then usual rule to calculate the transmission time: "The difference between these results [apparent Saint-Louis-to-Dakar and apparent Dakar-to-Saint-Louis], which is [0.326 seconds] represents double the time taken for the transmission of the electrical wave plus other errors."[40] Routinized as they were, such corrections were critical for the highly exacting measurements facing Poincaré and his colleagues at the Bureau. In measurement after measurement, compensating for time delay loomed large as surveyors struggled to eliminate the recalcitrant longitude clashes that still remained between Paris and Greenwich, or between Paris and the far-flung French colonies of West Africa, North Africa, and the Far East.

In a certain, unfairly retrospective sense, it is obvious that the longitude finders had to take into account such a time of transmission. After all, they claimed to be recording longitudes to the thousandth of a second, and the transmission time for light over 6,000 kilometers was about a fiftieth of a second. Electrical waves wound their way through equivalent lengths of copper undersea cable several times more slowly, making the correction nearly a *tenth* of a second. In the late-nineteenth-century world that vaunted cartographic precision, this was too much — every second of time error at the equator meant half a kilometer of east-west confusion.

Poincaré's remarks in his 1898 "The Measure of Time" about how to define simultaneity by telegraphic signal exchange were not, therefore, imaginary speculation about conventionality. Here was one of the three permanent Academy members of the French Bureau of Longitude, by far the most famous, just a few months before he was elected president, reporting on standard geodesic practice. All around him he could see simultaneity operationalized through the Bureau's active network of cables, pendula, and mobile observatories.

Poincaré could also see how scientific procedure crossed into philosophy. Some twelve years before, his Polytechnique friend, physicist-philosopher Auguste Calinon, had urged him toward a naturalized view of time and simultaneity. Poincaré had responded sympathetically. Then, in 1897—at precisely the moment Poincaré was most deeply immersed in the decimalization of time—Calinon published a new book. His thirty-paged tract, A *Study of the Various Quantities of Mathematics*, laid out the vicious circle of our reasoning about the equality of durations. A container is filled with water, then emptied through a spout at its base. Does the emptying take equally long each time the process is repeated? To answer the question presupposes an independent measure of time. But Calinon pointed out that the same question then would occur in this independent measure: what would calibrate the calibrator? Poincaré clearly was struck by Calinon's formulation and quoted it in "The Measure of Time": "One of the circumstances of this phenomenon [the time needed to empty the water from a container] is the rotation of the earth; if this speed of rotation varies, it constitutes, in its reproduction of the phenomenon, a circumstance that no longer remains the same. But to suppose this speed of rotation is constant is to suppose that one knows how to measure time."[41]

Calinon went even further than Poincaré indicated. He emphasized the historical arbitrariness of human time divisions; the seasons, for example, were chosen not based on any scientific or metaphysical conception, but for simple material utility. When scientists entered the picture, they simply chose the "simplest and most convenient" mechanism, which for Calinon meant that the

motion of the hands of a clock were such that, by their use, the formulae for the movement of the planets would be as "simple as possible." Calinon concluded that there was an irreducible *choice* in
the measure of time, one that had to be based on convenience: "In
reality, measurable duration is a variable, chosen from among all
the variables present in the study of movements, because it lends
itself particularly well to the expression of simple laws of movement."[42]

In the intersecting circles around Polytechnique, Poincaré
continually faced "choice," "convenience," and "simplicity" in the
measure of time, both throughout the technical world (railroaders,
electricians, astronomers) and in the scientific philosophy of his circle of Polytechnicians. His "Measure of Time" of January 1898 precisely marks that intersection. The measure of time is a convention,
one tied to the realities of scientific procedure. It is essential, however, to also see what the paper was *not*. Poincaré's "Measure of
Time" was *not* an inquiry into simultaneity that took this principled
correction of simultaneity into the heart of his physics. There is not
a single word in this article about frames of reference, electrodynamics, or Lorentz's theory of the electron. Like his field geodesists,
in early 1898 Poincaré saw electromagnetic exchange as key to providing a conventional, rule-governed approach to simultaneity. Also
like his geodesist colleagues, he saw the scientific consequence of
the time-delay definition of simultaneity as just another correction,
wedged between the inertial hesitation of an oscillating mirror and
the psychophysiology of the observers. Unlike the geodesists, Poincaré saw a philosophically significant point in that correction. Like
Calinon, Poincaré saw philosophy in the scientist's measurement of
time. But unlike Calinon, Poincaré had a direct involvement in the
coordination of distant clocks. Only Poincaré stood precisely at that
intersection; only he seized on the exchange of electrical signals to
make a routine physical process into the basis for a philosophical
redefinition of time and simultaneity. By crossing the electrical sur-

veyor's moves with those of the naturalizing philosopher, a piece of everyday technology suddenly functioned in both domains at once; it could serve in the clock room at Montsouris and grace the *Review of Metaphysics and Morals.*

In Poincaré's 1897 field of action, conventions of simultaneity were everywhere: no one would even bother to claim authorship of the longitude finder's rule for transmission-time correction. Below the threshold of signed patents or authored scientific papers, the geodesist's correction was part of the vast sea of anonymous knowledge that structured the telegraphic surveyor's everyday practice. More generally, conventions surged into the visible in every international technical conference, in every accord over length, electricity, telegraphy, meridians, and time.

Recall Poincaré's words about the nature of physical laws in the "Measure of Time": "no general rule, no rigorous rule, [rather] a multitude of little rules applicable to each particular case." Poincaré believed that the redefinition of time was another longitude hunter's correction that should not ultimately dent the simplicity of Newton's epochal laws: "These rules are not imposed on us, and one can amuse oneself by inventing others. Nonetheless we would not know how to deviate from these rules without greatly complicating the formulation of the laws of physics, of mechanics, of astronomy." We choose these rules, Poincaré insisted in oft-cited words, not because they are true, but because they are convenient. As far as Poincaré was concerned in 1898, that meant guarding the long-established laws of Newton for which a simpler alternative simply could not be imagined. "In other words," he concluded, "all these rules, all these definitions, are nothing but the fruit of an unconscious opportunism." In Poincaré's insistence on the conventionality of time, we hear ideas that we can now recognize as having echoed through the halls and wires of the Bureau of Longitude and Ecole Polytechnique, a technical universe in which diplomats, scientists, and engineers used international conventions to manage the colliding imperial networks of space, time, telegraphs, and maps. This world too was Poincaré's, a way of being in science that stood a long way from Einstein's.

Mission to Quito

Poincaré's—and the Bureau's—engagement with telegraphic longitude work intensified between 1898 and 1900. Within a few months of the Bureau's publication of their longitude fix of Dakar and Saint-Louis, the Bureau shifted into high gear to launch another major expedition, this time to Ecuador, with Poincaré as scientific secretary. Discussion of such an expedition had begun long before, certainly as far back as 1889 at the Exposition Universelle. In the midst of that techno-fair, the American delegate to the International Geodesic Association had asked for a determination of the length of a meridial arc (an arc along a line of longitude) as part of an effort to determine the shape of the earth. If the earth widened around the equator and flattened at the poles (as Newton had predicted two centuries earlier), then a longitudinal arc covering, say, 5 degrees of astronomical latitude near the equator should be shorter than a 5-degree arc nearer to one of the poles. That the earth was oblate was long known: the question now was to determine that shape with a precision fitting the age. The President of Ecuador approved, telling the French that they would have the full help of his government. As the proposal bounced between the Ministry of War, the Ministry of Foreign Affairs, and the Army's Geographical Service, the cost began to rise, even as its importance remained clear.[43] Nor did Ecuadorian politics hold still. With his ascension to Ecuador's presidency on 5 June 1895, General Eloy Alfaro began the secular Liberal Revolution that aspired to reform every sector of national life. The bulging earth had to wait.

On 7 October 1898, the American delegate at that year's International Geodesic Association meeting at Stuttgart again pleaded for a long-overdue remeasurement of the shape of the earth by way of an expedition to Quito, Ecuador. Bouquet de la Grye, the French delegate, reminded the assembled that, just four years previously, the Plenipotentiary Minister from Ecuador had proposed such an expedition but that the 1895 revolution had required an "adjournment" of the question. After the American delegate's polite acknowledgment, the British representative spoke less diffidently: wishes were fine, but the geodesists needed to say precisely what could and should be done. London demanded France's response to the need for meridial action. De la Grye was "persuaded that in France one would look with a favorable eye on the wish that we reopen negotiations on a new measurement of the arc of Peru." Having watched American surveying forces nail locations up and down the Americas, and witnessing the British near-monopoly on undersea cables in pursuit of their own cartographic ambitions, the French understood full well that, if they hesitated, the American Coast Survey would grab the job. But if the Americans seized the map, as Poincaré put it, "the honor of our country would be ravished."[44] France mobilized.

From Paris the Minister of Public Instruction offered Fr. 20,000 for a reconnaissance mission to Quito using personnel assigned by the Minister of War. In this up-to-date foray, the Minister of Public Instruction would extend the much-heralded eighteenth-century measurement by 1 degree of latitude to the north and 2 degrees to the south, with the area around each base to be given topographical summaries and surveys of the horizon. Immediately, the French would send an avant-garde expedition that would move fast, traversing 3,500 kilometers of some of the highest mountains in the world in only four months. Departing from Bordeaux on 26 May 1899, the team worked at breakneck speed. While they were away, tensions with Britain mounted; there was the affair of the French cable lines that had "inexplicably" failed during the French-English race to stake colonial lands on the Nile, and then the outright British censorship of cables on 17 November 1899 during the Boer War. On 8 December 1899, the French Minister of Colonies announced a cable bill of unprecedented size. Just a week later, the Council of Ministers passed a plan for an imperial network budgeted at 100 million francs. Polemics over the cable project appeared in publications on both right and left.[45] The Quito mapping team sailed back to France in the midst of this brouhaha, arriving in Paris on the last day of the nineteenth century.

At their meeting on Monday, 23 July 1900, it was up to the French Academy of Sciences to decide whether and how to undertake the full mission. Poincaré left little doubt as to the obligations he expected his country's scientists to meet:

> If our country owes itself its part in the conquests of modern science, then all the more reason we must not abandon a position on which our fathers hoisted, so to speak, the intellectual flag of France. Our rights have been publicly recognized. Should we respond to these courteous invitations by a declaration of impotence? France is as dynamic and richer than it was one hundred and fifty years ago. Why would she leave to younger nations the task of completing what the France of yesteryear had begun?[46]

The minister asked the Academy to control the operations; Poincaré urged the august body to go further. Shouldn't they follow the legacy of their intrepid eighteenth-century forebears, Louis Godin, Pierre Bouguer, and Charles-Marie La Condamine, who had surveyed a degree of arc in Quito beginning in April of 1735? A century and a half later, however, academicians could well ask if it was necessary for scientists to accompany the expedition into the Andes. Such a voyage would demand more than calculational abilities, as the rather sedentary Poincaré conceded:

A high scientific competence, technical facility, and habits of scrupulous regularity would be indispensable, but would not suffice. One must be in a state that would support great exhaustion, in territory without resources and in every climate. One must know how to lead men, obtain obedience of one's collaborators and impose that obedience on the half-civilized servants one is forced to employ. All these intellectual, physical, and moral qualities are found united in the officers of our geographical service.[47]

What the academy could do from the safety of Paris was to maintain scientific control by a commission charged with examining the observation notebooks.

Poincaré himself headed mission control. He made it clear in his report of 25 July 1900 that imposing the grid on this equatorial region would demand three bases. To ensure reliability, these should be located using the same instruments employed in determining meridian length in France. Quito itself would be manned by a most gifted (French) astronomer, François Gonnessiat of the Lyon Observatory, with the whole operation generously funded by an anonymous donor later revealed to be Prince Roland Bonaparte. With French officers operating the extreme stations, Gonnessiat would make simultaneous observations at Quito, ensuring accurate longitude fixes. As always, the telegraph was key in the establishment of distant simultaneity: one wire would run from Quito to Gayaquil, boosting its signal with a relay; another would tie Quito to the distant north station. If the relays introduced unacceptable error into the simultaneity determination, as they often did, Poincaré was prepared to slice them from the system by boosting the battery power of the initial station. At present, the team was already working to string the telegraph network all the way to the southern station. So the extreme ends of this far-flung mountain network would be tied back to Quito, with Quito's time bound to Gayaquil's. Gayaquil, in turn, would coordinate its

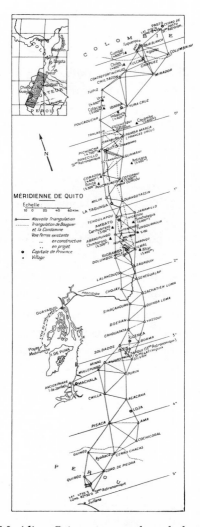

Figure 4.3 Quito Meridian. *Poincaré reported regularly on the Quito expedition. In this picture, the multiple triangulations are shown; these were used to measure the length of a longitudinal arc in order to give greater precision to the shape of the earth. This network of latitude and longitude measurements was then linked by cable to the world telegraph net—and would be calibrated to the French prime meridian.* SOURCE: POINCARÉ, *OEUVRES COMPLÈTES*, VOL. 5, P. 575.

clocks through undersea cables tied to the general world network via North America.[48]

The academy approved Poincaré's report as did the International Geodesic Association in October 1900, with the American delegate congratulating the French and reminding them that the United States would like to be asked first if any additional help were needed. Not likely. On 9 December 1900, Captains Maurain (of the engineers) and Lallemand (of the artillery) left for Ecuador as the avant-garde of what they expected would be a four-year mission to Quito.[49]

On Monday, 28 April 1902, Poincaré reported to the Academicians in Paris: Members of the advanced team had spent months buying pack animals and setting up their convoys. By the time they started up into the mountains, the French expeditionary force had 120 mules and were accompanied by forty Indian porters and six Ecuadorian officers. The vital measuring stick, consisting of two different metals with different responses to heat, had to be carried on a man's back. At night the humidity so distorted the micrometer threads that they used for sighting that observations became almost impossible between six and nine in the morning. Then, at 11:00 A.M., howling winds blasted dust through every crevice in the campsite, wrecking instruments and torturing the explorers. Wild temperature changes plunged through the camp so quickly that the team wondered if their measuring devices would ever settle into a reliable state of equilibrium. Amidst all this, the French decided they would prefer to use their personal equations rather than switch positions of the observers, which would have left Indians, alone, in charge of the base camp. Lallemand took two men on a side mission to the far north in the hopes of extending the network into Colombia, but they were promptly stymied by political turbulence. In a somewhat optimistic aside, Poincaré conveyed the expedition leaders' hope that, before 1904, they would have established telegraphic simultaneity to all the far north stations short of Colombia.[50]

A year later, Poincaré was back before the Academy, once again having drafted a report on the hard-driving expedition. It was Monday, 6 April 1903, and the commissioners noted with regret that the previous year's progress had been held up on almost every front. First the summits were almost constantly lost in fog, making sightings impossible. Lieutenant Perrier had spent three months at his assigned post in Mirador, 12,000 feet above sea level, almost constantly blanketed in clouds. Incessant rain pounded on his site, the visible horizon limited to the camp itself. Furious winds shook everything in the barracks. In one period of fifteen days, he was able to make but one of his twenty-one measurements and caught not a single glimpse of the signal he awaited from Yura Cruz. The valleys separating them flowed with a rushing river of clouds from the East. After months of isolation, Perrier's perseverance was rewarded with calmer, clearer weather. In an intense burst of activity, the lieutenant completed his mission.[51]

Other brigades faced similar problems and, by Poincaré's reckoning, gave the same proof of their quality. At Tacunga, Captain Maurain could only observe intermittently, seizing upon occasional clearing. A violent east wind accompanied by snow squalls made his work exceedingly painful. Gusts tore roof braces from the observation station, ripping up tents right and left. At Cahuito station Lacombe found himself stranded for days in the fog and snow, unable to execute a single observation. And Lallemand, who directed the reconnaissance brigade and signal construction across exceedingly tough terrain, fell into a crevasse in Cotopaxi. Other soldiers recovered him, but for three weeks he was laid low. Amidst these battles with the elements, the team speculated that the ferocious weather might be linked to the volcanic activity following a catastrophic eruption on Martinique.[52]

But not all the misfortunes of the longitude men could be laid at the feet of nature. Both indigenous whites and Indians tore into the geodesists' painfully established reference points. Apparently surveying rods signaled more than lat-long positions. For locals, these markers promised buried wealth. In their search for gold, not only did the treasure hunters remove rods, they dug up everything remotely near them. So the French—after their months in the oxygen-deprived, snow-driven peaks—had to resignal, remeasure, and re-mark the sites, sometimes two or even three times. Enlisting the Church, the French enjoined local bishops and cures to stave off the inhabitants, but neither science nor God deterred the diggers. Over the course of 1904, the French found their stations destroyed some eighteen times, forcing the remeasurement of 360 pairs of points in horrendous high-mountain conditions. The following year, to their dismay, the simultaneity team realized, while freezing in the craggy heights, that their native informants' reports on visibility had been, as they put it, "not exact." The geographers finally understood that their informants had never voluntarily gone anywhere near such heights. On those few occasions that the locals had actually ventured to high altitudes, "visibility" meant being able to see enough to climb and descend. Unlike the forces of simultaneity, the native people had neither a need nor a desire to see far enough to flash latitude and longitude. Meanwhile Lallemand contracted yellow fever; his colleagues sent him back to France.

Poincaré delivered this encomium for the mission: "The long days of waiting in snow and fog did not lead to an instant of discouragement and the zeal, the constancy, and the devotion of these officers and of all the personnel ought never be denied. There are grounds for congratulating these valiant pioneers of Science for their courage and the results they obtained."[53] Map coordinates in hand, after eight years of effort to fix time and space, Poincaré reported in 1907 that the expedition had returned to France.

Etherial Time

In following the Quito expedition all the way back to Paris, we have gotten ahead of ourselves. For during the entire span of the longitude expedition—from planning in 1899 to its conclusion in 1907—Poincaré pressed the technology of simultaneity in two other, very different domains: philosophy and physics. This was no accident. From the time he completed his report on the Quito mission's objectives (25 July 1900), Poincaré was fully immersed in the details of telegraphic simultaneity, down to the battery power of the telegraph lines. He had made the case to his colleagues that mapping Quito was of double importance—as a defense of French honor ("hoisting the intellectual flag of France") and as a technical-scientific problem (determining the shape of the earth, mapping the world). But that was not all.

No longer thinking purely mathematically, Poincaré now reflected on the philosophical, conventional basis of physics more generally. Just a few days after urging the Quito project on his fellow Academicians, he delivered a paper at a major philosophy conference (2 August 1900),[54] where he asked a fundamental question about what he considered *the* most central science: could the basic ideas of mechanics itself be altered? Among his French compatriots, Poincaré argued, mechanics had long been treated as beyond the reach of experience, as a deductive science that would inevitably lead to conclusions from assumed first principles. By contrast, British mechanics was at root not a theoretical but an experimental science. If there was going to be any real progress in mechanics, this cross-channel clash would have to be sorted out analytically. What part of this most exalted science was experimental? What part mathematical? And—as he put it—what part conventional?

In response, Poincaré laid out for the philosophers his judgment of what we knew about the starting points of this most fundamental of sciences, drawing from the full range of his work on the foundations of geometry, geodesy, physics, and philosophy. In his words:

1. There is no absolute space, and we only conceive of relative motion; and yet in most cases mechanical facts are enunciated as if there is an absolute space to which they can be referred.

2. There is no absolute time. When we say that two periods are equal, the statement has no meaning, and can only acquire a meaning by a convention.

3. Not only have we no direct intuition of the equality of two periods, but we have not even direct intuition of the simultaneity of two events occurring in two different places. I have explained this in an article entitled "Mesure du Temps."

4. Finally, is not Euclidean geometry in itself only a kind of convention of language?

For Poincaré, mechanics could be alternately enunciated using a non-Euclidean geometry. It might be less convenient than mechanics articulated using ordinary Euclidean geometry, but it would be altogether as legitimate. Absolute time, absolute space, absolute simultaneity, and even absolute (Euclidean) geometry were not *imposed* on mechanics. Such absolutes "no more existed before mechanics than the French language can be logically said to have existed before the truths which are expressed in French."[55]

These four summary statements captured a great deal of Poincaré's approach. The first reemphasized his philosophical-physical objection to absolute space. The second and third recapped his 1898 "Measure of Time," and the fourth linked the discussion to his older work on the conventionality of geometry. "Are the laws of acceleration and of the composition of forces only arbitrary conventions? Conventions, yes; arbitrary, no—they would be so if we lost sight of the experiments which led the founders of the science to adopt them, and which, imperfect as they were, were sufficient to justify their adoption. It is well from time to time to let our attention dwell on the experimental origin of these conventions." Experiments, for Poincaré, were the raw materials of physics out of which theory aimed to produce the greatest number of predictions with the highest probability. That is, experiments could serve as the "basis" for mechanics by suggesting, as it were, the principles and rough behavior of the world that we want embodied in the conventions (definitions) of concepts such as force, mass, and acceleration. But that is not to say that experiments can simply invalidate starting principles; at worst, in his view, experiments would reveal that a fundamental law was only approximately true, "and we know that already."[56] Poincaré summarized his reflections on the role of theory when he spoke to the physicists (also in 1900). As so often, he invoked the machine-based factory, framing experiments as the "raw materials" and theory as the organizing guide: "The problem is, so to speak, to increase the output of the scientific machine."[57]

Even in mathematics, machines and mechanomorphic structures were vital for Poincaré. Back in 1889, Poincaré had laid into the logicist purveyors of "teratological" functions. He came back to the theme in August 1900, this time while addressing the mathematicians gathered for their international congress in Paris. Again he pitted the logicists against the intuitionists. While he judged both important for the development of mathematics, there was no doubt on which side he stood. One mathematician (a logician, in Poincaré's division) could cover pages and pages of print in order to demonstrate unequivocally that an angle could be divided into any number of equal parts. By contrast, Poincaré called his audience's attention to the Göttingen mathematician Felix Klein: "He is studying one of the most abstract questions of the theory of functions: to determine whether on a given [abstract mathematical] Riemann surface there always exists a function admitting of given singularities [roughly: points where the function becomes infinite]. What does the celebrated German geometer do? He replaces his Riemann surface with a metallic surface whose electric conductivity varies according to certain laws. He connects two of its points with the two poles of a battery. The current, says he, must pass, and the distribution of this current on the surface will define a function whose singularities will be precisely those called for." Now Klein knew perfectly well that this reasoning was not rigorous, but (says

Poincaré) "he finds in it, if not a rigorous demonstration, at least a kind of moral certainty. A logician would have rejected with horror such a conception." More precisely, the logician would never have been able to formulate the intuitionist's thought.[58] But formulate such machine thoughts Poincaré certainly did: in pure mathematics, in geodesy, in philosophy. In the midst of it all, he pressed ahead with studies of electricity and magnetism, bringing his ideas about electromagnetic clock coordination into the heartland of physics itself.

On 10 December 1900, just two days after Poincaré had sent Lallemand and Maurain on their Quito telegraphic longitude mission from the Gare Saint-Nazaire, Poincaré stood at the lectern in the University of Leiden, outlining his ideas on space, time, and ether. He and other luminaries had gathered to honor H. A. Lorentz. Lorentz was for Poincaré (and, I should add, for Einstein) a singular figure among physicists. In many ways, it was Lorentz who had helped exemplify the new professional category, "theoretical physicist." For many physicists, especially those outside Britain, it was Lorentz who had put Maxwell's theory of electricity and magnetism into a comprehensible form. Instead of following the British tradition of reducing all matter to etherial flows, vortices, stresses, and strains, Lorentz had a starker doctrine: there were *two* kinds of things in the world, electric and magnetic fields (states of the ether) on the one hand; and material, charged particles moving through the ether, on the other. Fields could act on particles, particles could create fields. But Lorentz had done still more, as he struggled to explain the seeming failure of experimentalists to reveal the motion of objects—including the earth—through the vast ether that was supposed to pervade the Universe.

For some time Poincaré had, like all other up-to-date physicists, been thoroughly familiar with Lorentz's theory. He had also been respectfully critical of it when he lectured on it at the Sorbonne in 1899, believing it to be the best available theory. Even with Lorentz seated in the Leiden audience in 1900, Poincaré said, it disturbed him that Lorentz's account violated the principle of action and reaction. A cannon retreats immediately when it shoots a cannonball, but the ether and atoms had the odd property (according to Lorentz) that an atom would act on the ether but that the material receiver of the atom's action would only react later. What happened in the meantime? The ether seemed too insubstantial to carry momentum. Poincaré told Lorentz and the assembled that he could mitigate this objection. "But I disdain that excuse because I've got one a hundred times better, *Good theories are flexible.* . . . [I]f a theory reveals to us certain true relations, it can dress itself in a thousand diverse forms, it will resist all assaults and that which is its essence will not change." Lorentz had created one of those flexible theories, one of the truly good ones: "I will not apologize" for criticizing the theory, he said. Instead, he only regretted that he had so little to add to Lorentz's old ideas.[59] Despite the disclaimer, Poincaré went on to transform the physical significance of Lorentz's theory.

At the root of Lorentz's old ideas had been a theoretical account of a seemingly intractable difficulty in ether physics. Was the ether simply dragged along by transparent matter? If ether was hauled along by the earth's atmosphere, that could explain why experiments never seemed to reveal earthly motion through the ether. Unfortunately, a mid-nineteenth-century experiment by the French physicist Armand-Hippolyte Fizeau seemed to rule that out. By shining light through water flowing at different speeds, he could demonstrate that if the ether was dragged, it was only very partially. But if the ether was *not* dragged by matter (and so was fixed once and for all in the Universe), then we ought to be able to detect our motion through it. This is just what the American experimenter, Albert Michelson, set out to uncover by using an extraordinarily precise optical "interferometer" that he had invented (see figure 4.4).

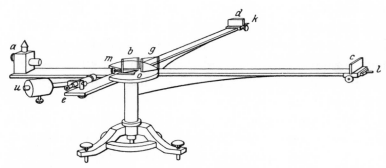

Figure 4.4 Hunting the Ether. With a remarkable series of experiments, Albert Michelson sought to measure the earth's motion through the elusive ether. In the 1881 device shown here, he launched a beam of light from a that was split by a half-silvered mirror at b: one-half of the ray reflected off d and into the eyepiece e. The other half of the ray penetrated the mirror at b, reflected from c, and was then bounced from b to the eyepiece e. At the eyepiece the two rays interfered with each other, showing the observer a characteristic pattern of light and dark. If one wave was delayed—by so little as a part of a wavelength of light—this pattern would visibly shift. So if the earth really was flying through the ether, then the "ether wind" would affect the relative time it took for the two beams to make their round-trips (the relative phase of the two waves would shift). Consequently, Michelson fully expected that if he rotated the apparatus, he would see a change in the interference patterns of the two rays. But no matter how he twisted his staggeringly sensitive instrument, the dark and light patterns did not budge. To Lorentz and Poincaré, this meant that the interferometer arms—like all matter—were contracted by their rush through the ether in just such a way as to hide the effect of the ether. To Einstein it was one more suggestive piece of evidence that the very idea of the ether was "superfluous." SOURCE: MICHELSON, "THE RELATIVE MOTION OF THE EARTH AND LUMINIFEROUS ETHER," *AMERICAN JOURNAL OF SCIENCE*, 3RD SERIES, VOL. XXII, NO. 128 (AUGUST 1881), P. 124.

Indeed, Michelson thought he had the ether cornered. For if there truly was an ether wind, then the round-trip time for a light beam should change, depending on whether the light was being sent across or into the wind. But rotating the apparatus showed not the slightest twitch of the interference dark spot; to an extraordinary degree of accuracy, it seemed that no motion through the ether could be detected by optical means. Though Michelson saw his efforts to find the ether as a dismal failure, other physicists, including Lorentz, were moved to theorize.

Taking Michelson's null result into account, in 1892 Lorentz assumed the existence of a static ether, and introduced the startling notion that any object moving through the ether contracted in its direction of travel. Bizarre as this "Lorentz contraction" sounded, the gambit worked, in the sense that a judicious choice of this contraction factor exactly compensated for the effect of the putative ether wind. Lorentz's contraction explained why high-precision experiments—even the extraordinary Michelson interferometer— would be powerless to uncover the effects of the etherial breeze on optical phenomena.

Remarkably, Lorentz's contraction hypothesis was not enough. In the course of demonstrating that *all* optical phenomena could be described in approximately the same way, he introduced in 1895 a second innovation, a fictional "local time."[60] Lorentz's idea was that there was one true physical time, t_{true}. True time was the appropriate time to use for objects at rest in the ether. For any object moving in the ether, it proved useful for Lorentz to introduce this fictional time (a mathematical artifice) in terms of which the laws of electricity and magnetism for that object would artificially resemble the laws for an object sitting still in the ether. This helpful quantity (t_{local}) depended on the speed (v) of the object plowing through the ether, the speed of light (c), and the location of the object (x); t_{local} was just t_{true} minus vx/c^2. Why did Lorentz choose this local time? Only because it gave a sharp, if purely formal, result: local time allowed a real object moving in the ether to be redescribed as a fictional object at rest in the ether. Local time for Lorentz was but a mathematical convenience.

Poincaré had viewed Lorentz's theory and its assumptions about

length contraction and "local time" with reserve. Even in his Sorbonne course of 1899, he did not connect local time to his
technological-philosophical definition of simultaneity; in fact he
remarked that corrections due to Lorentz's "local time" were so
small for the earth moving through the ether (the correction was
only three-billionths of a second for two clocks separated by a kilometer) that he would simply ignore the correction.[61] Meanwhile,
Poincaré's immersion in telegraphic longitude determination intensified. Not only was he in the thick of planning for the Quito expedition during 1899–1900, he was, in 1899, also president of the
Bureau of Longitude. If he had somehow stood back from the
details of the simultaneity procedures before, now he was fully
engaged.

On 23 June 1899, for example, Poincaré wrote to the Royal
Astronomer about the confounding discrepancy between the
French and British measurements of the Paris-Greenwich longitudinal difference. Could the British begin a new effort immediately?
William Christie responded 3 August pleading for more time, but
promising he would prepare detailed, published analyses of their
procedures and instruments, noting that "it is desirable too that the
French results for the longitude Paris-Greenwich should also be
published in detail." Christie copied other letters to Poincaré about
the problem, letters Christie had penned to Poincaré's Bureau colleagues. To Loewy, Christie had speculated that the error source
might be a discrepancy between leveling the star-sighting instrument based on a plum line and on a spirit level. Apparently the
French Army had implied that the fault might lie with British
clocks; no, Christie replied, not possible. Instead he suggested that
both Greenwich and Paris repeat the Paris-Greenwich procedure,
only now between two piers inside Greenwich and again inside
Montsouris. Presumably that would isolate any error accrued during the signal transmission under the English Channel. Poincaré
replied immediately (9 August 1899) that he looked forward to

receiving the British publications as soon as possible and that he hoped that they would contain the detailed calculations and the methods by which the data were reduced; in return, the French would open their books as both sides urgently tried to cut the embarrassing longitude discrepancy between Paris and Greenwich.[62]

In the midst of these longitudinal measurements and troubleshooting, Poincaré came face-to-face with Lorentz at the December 1900 meeting. Preparing for the event, Poincaré reinterpreted, in a strikingly new fashion, Lorentz's "local time." First (though he did not cite it) Poincaré introduced his 1898 "Measure of Time" assertion that simultaneity could be given between clocks synchronized by the exchange of electromagnetic signals. That was the argument he had produced at the intersection of longitude and metaphysics. Now he went further, moving the technological-philosophical idea of electrically coordinated clocks into physics itself, forming a *triple* intersection. For the first time he pursued his clock synchronization method for clocks moving through the ether. Poincaré's sudden understanding: when executed while moving through the ether, the telegrapher's procedure for synchronizing clocks gave Lorentz's fictional local time, t_{local}.

"*Local time*," Poincaré insisted, *was just the "time" clocks showed in a moving reference frame when they were coordinated by sending an electromagnetic signal from one to the others*. No mathematical fiction, this was what *moving observers* would actually *see*:

I suppose that observers placed in different points set their watches by means of optical signals; that they try to correct these signals by the transmission time, but that, ignoring their translatory motion and consequently believing that the signals travel at the same speed in both directions, they limit themselves to crossing the observations, by sending one signal from A to B, then another from B to A. The local time t' is the time indicated by watches set in this manner.[63]

Here is the electrical simultaneity procedure that we know twice: from Poincaré longitude-finder and from Poincaré-philosopher. Poincaré-physicist inserts the procedure into a *moving* frame of reference.

Here, more precisely, is Poincaré's argument. Consider a moving frame containing clocks A and B that is moving to the right through the ether at a constant speed v. Because of this motion, the light signal from A to B (say) encounters a headwind as it beats *into* the ether wind; the signal speed is the speed of light c, minus the wind speed, v (like an airplane encountering a headwind as it flies from Europe to the United States). The return light signal from B to A *boosts* its speed by an ether tailwind: its speed is therefore light speed *plus* ether wind speed, $(c + v)$.

A → (light velocity: $c - v$) →B	///// Ether (at rest) /////
A← (light velocity: $c + v$) ← B	—→ v = *velocity* of frame →

Because speeds differ in the two directions, a naive one-way clock coordination conducted by moving physicists would go wrong. For example, if B sets her clock using a signal from A, she is using a signal that is "really" going at $c + v$ (relative to the ether). B gets the signal too soon relative to a "correct" signal that would travel at c. The further away B is from A, the bigger the time discrepancy between the moment she gets the tailwind-boosted signal and the moment she would have received it were it sent at the stately speed of c. So B needs to set her clock back a bit—less if she is close to A, more if she is farther away. Poincaré observed that the offset correction to account for the too-speedy signal ($-vx/c^2$) yielded just the fictional "local time" correction of Lorentz's "local time."[64] Poincaré's message: Clocks moving in the ether must be coordinated by the launch of electromagnetic signals, as before. But coordination in a moving frame demands an offset of the clocks to compensate for the effect of the ether wind.

Lorentz acknowledged the force of Poincaré's criticisms in a letter he posted on 20 January 1901—not a word, however, about Poincaré's interpretation of "local time." On other matters Lorentz readily ceded: there was a problem with the principle of reaction, and as far as Lorentz could tell that might forever be the case if physicists wanted to account for the experiments; if the ether was absolutely rigid and immovable, then one could not coherently talk about forces acting on it. So ether, as Lorentz had long argued, could exert force on electrons, but electrons could not exert force on the ether. Nor could one part of the ether act on another—any attempt to speak this way was, in effect, invoking "mathematical fictions." No doubt such fictions were useful for calculating the ways in which the ether eventually acted to move electrons. But they were fictions nonetheless. Alternatively, Lorentz speculated, it might be possible to say that the ether was infinitely massive, in which case electrons could act on it without putting it into motion: "But that way out seems to me pretty artificial."

Lorentz had produced an extraordinary theory, one that had utterly transformed physics by dividing the world between a vast unmoveable ether and material electrons. Immensely successful, the theory accounted for a myriad of experiments from spectral lines to the explanation of simple optics like reflection. But as Lorentz struggled to extend the theory, he found himself reaching for a variety of tools that he readily allowed were artificial. Lorentz supposed that matter contracts as it plunges through the ether, that the principle of action and reaction is violated, and that the theory needed the mathematical fiction of "local time."

Poincaré approached Lorentz's theory differently. As always when treating theories, Poincaré's procedure was to isolate the building blocks of the theory and then to manipulate the theory's most useful parts in order to advance. Here, at Leiden in 1900, he joined Lorentz's local time with his own conventional interpretation of simultaneity—the longitude finder's and philosopher's convention. Without fanfare, he showed how Lorentz's local time could be physically interpreted—the telegrapher's convention now set in an etherial wind.

None of Poincaré's alterations of the theory sullied his assessment of Lorentz. On the contrary, in January 1902, Poincaré nominated Lorentz for the Nobel Prize, explaining to the Swedish authorities how Lorentz had seized the failure of previous physicists to find the ether or explain their failure to do so. Poincaré: "It was evident that there must have been a general reason; Mr. Lorentz discovered that reason and put it in a striking form by his ingenious invention of 'reduced time'." Two phenomena that occurred in different places could appear to be simultaneous even when they were not. Everything happened as if the clock in one of the places was running behind the other, and as if no conceivable experience could allow the discovery of this discordance. As Poincaré saw it, the failure of experimentalists to observe the movement of the earth through the ether was just one such futile attempt.[65]

Continuing his éloge to the Nobel committee, Poincaré acknowledged that people often judge theories such as Lorentz's as fragile. The sight of the ruins of past theorizing may make anyone into a skeptic, he allowed. But if Lorentz's theories were to join others in the vast cemetery of defeated accounts of the world, no one could justifiably claim that Lorentz had randomly predicted true facts (*faits vrais*). "No, it is not by chance, it is because [Lorentz's theory] revealed to us relations unknown until then between facts that apparently were strangers one to the other, and that these relations are real; and that they would be so even if electrons didn't exist. It is that sort of truth that one can hope to find in a theory, and these truths will survive beyond the theory. It is because we believe

that the work of Lorentz contains many such truths that we propose
that he be rewarded for them."[66]

A *Triple Conjunction*

By the end of 1902, Poincaré had spent a full decade facing the
problem of time coordination from three very different perspectives.
As one of the exalted Bureau of Longitude's Academy members
since January 1893, Poincaré had helped lead the institution in its
quest to cover the world with synchronized time. When the issue of
conventionally restructuring time into a decimal system arose in
earnest during the mid-1890s, it was Poincaré who had directed the
effort to evaluate the alternatives, culminating in the 1897 report.
Then back to philosophy: in 1898 he proclaimed to a primarily
philosophical audience that simultaneity was nothing other than a
convention, a convention that would define simultaneity precisely
as his Bureau had done with their telegraphic coordination of
clocks. From the *Revue de Metaphysique et de Morale*, it was but a
few months before Poincaré was back, deeper than ever, in expeditionary longitudinism. From 1899 forward he had been the liaison
between the Academy and the complex and dangerous longitude
mission to Quito while the mission struggled to bind time and geography through telegraphically flashed simultaneity. That summer
of 1900 he issued his strongest-yet philosophical statement of the
conventionality of simultaneity, a statement that appeared in one of
the most widely cited sections of *Science and Hypothesis*. Then a
return to physics. In December 1900, Poincaré "reviewed"
Lorentz's theory, turning Lorentz's mathematical fiction "local
time" into a telegrapher's procedure, where observers moving
through the ether synchronized clocks by exchanging signals.

This was not merely a matter of "generalizing" from the physics
to the philosophy or "applying" abstract notions from mathematics
or philosophy to physics. Instead, Poincaré was working out a new

concept of time, showing how it fit into the rules of three different games: geodesy, philosophy, and physics. Just because the statement of synchronization and simultaneity functioned within all three of Poincaré's worlds, it took on a singular importance.

All through Poincaré's work lay a Third Republic progressivism, a sense that all aspects of the world could be improved through an engaged reason, a rational, machinic modernism. His faith carried the optimism of an abstract engineering, an undying belief that it was within the reach of reason to get at any aspect of the map, whether of the geographical world or that of science. All things being equal, Poincaré would just as soon resolve a problem the way he handled the decimalization dispute—by setting the various positions against one another and cooly picking out the simplest (most convenient) relations among terms. *Relations* were what was real to Poincaré (the relations captured by Newton or Lorentz or by mathematics), not the particular objects of our senses. "It may be said . . . that the ether is no less real than any external body; to say this body exists is to say there is between the color of this body, its taste, its smell, an intimate bond, solid and persistent; to say the ether exists is to say there is a natural kinship between all the optical phenomena, and neither of the two propositions has less value than the other."[67] Objective reality was nothing other than the commonly held relationships among the phenomena of the world. There was no *otherworldly* plane of existence for Poincaré. The importance of scientific knowledge lay in the persistence of particular true relations, not in a back-of-the-curtain reality of Platonic forms or ungraspable noumena.

Poincaré generally shied away from overt political reference or moral absolutes. Still, once or twice, a different tenor breaks the neutral, moderated tone of his commentaries. We know, he told the General Association of Students in May 1903 in a Parisian restaurant, that there are twin fields on which action has to be taken: toward scientific truth and toward moral truth. "It seems that there is an antinomy between our two dearest aspirations, we want to be

sincere and serve the truth; we want to be strong and capable of acting. . . . From here the violence of the passions raised by a recent
affair. On both sides, most were driven by noble intentions." Allud-
ing to the Dreyfus Affair then wracking the country, Poincaré's ano-
dyne summary muted his own torn responses —*his* powerful
devotion to the French army and *his* equally urgent allegiance to
standards of demonstration. On 4 September 1899, in the midst of
Alfred Dreyfus's second trial, Poincaré had intervened by proxy. His
letter slammed the scientific basis of the charge that it was Dreyfus
who had written an incriminating torn sheet of paper (the famous
bordereau) promising French state secrets to a German military
attaché: "It is impossible," Poincaré concluded, that anyone "with
a solid scientific education" could give credence to the nonsensical
use of statistics presented by the prosecution. Despite that stark
judgment, the court reconvicted Dreyfus, though the President of
the Republic issued a pardon.[68] (Poincaré defended Dreyfus again
a few years later on technical grounds, though stronger ones.)
Return now to Poincaré's 1903 address to the assembled students.
Action had to join thought, he insisted.

> The day when France has no more soldiers, but only thinkers, Wil-
> helm II will be the master of Europe. Do you think then that
> Wilhelm II would have the same aspirations as you? Do you count
> on him using his power to defend your ideal? Or do you put your
> confidence in the people, and hope that they will commune in the
> same ideal? That's what one hoped in 1869. Do not imagine that
> what the Germans call right or liberty is the same thing that we
> call by the same names. . . . To forget our country is to betray the
> ideal and the truth. Without the soldiers of Year II, what would
> have remained of the Revolution?[69]

Poincaré added that every generation wonders about the fate of its
work, none more than his. "Cruelly hit at the moment they arrived

at manhood, my contemporaries set to work to repair the disaster [of the Franco-Prussian War]. . . . Years passed and deliverance did not come. And so we ask ourselves, have you inherited this dream, without which all our sacrifices would be in vain? Maybe . . . that which for us appeared as an intolerable injustice, . . . a bleeding wound is, for you, just a bad historical memory, like the distant disasters of Agincourt or Pavia."[70]

Crises and recuperation were always in view for Poincaré: in politics, but also in philosophy and in science. If his generation's political ideal was to "repair the disaster" of 1871, there were nearer opportunities where the scientific machine could re-arrange, repair, improve. In April 1904 he had a second opportunity to repair the damage caused by the Dreyfus crisis. Asked by the court to address the status of the famous *bordereau*, Poincaré, with two scientist-colleagues at the Observatory, scrutinized the evidence. They remeasured the handwriting by using precision astronomical instruments and recalculated every last bit of the prosecution's probabilistic reasoning designed to show the statistical near-certainty that the *bordereau* was Dreyfus's. Poincaré's conclusion: the handwriting analysis was nothing but the illegitimate application of badly reasoned probability on an incorrectly reconstituted document.[71] Backed by Poincaré's 100-page report of 2 August 1904, the intervention succeeded. Poincaré crushed, for example, the prosecution's claim that Dreyfus had written the word "*intérêt*" on the *bordereau*. Alphonse Bertillon, the prosecution's star handwriting expert, had contended that the word could only have been written using the grid of a military map—a scale that famously set the diameter of a "sou" equal to a kilometer. Along with his mathematician colleagues, Poincaré created an enormous enlargement of that single word, mapping it, as it were. Their conclusion: Bertillon's claims about the curvature, length, axis, and height of the letters were as arbitrary (changing for each letter) as his spurious microcaligraphic analysis of the circumflex accent. There was no "geometrical particularity" to this word. Nothing at all implied it was written by someone working at a staff officer's desk using military cartography.[72] Impressed, the court exculpated Dreyfus. Once again, Poincaré had mapped a technical fix to a profound crisis. This time it had taken more than a chart of coefficients (Poincaré's ploy to resolve the crisis of decimalization). But in a sense, the spirit was the same: reasoning through machines and calculations, he helped defuse a crisis.

Crises arose elsewhere. In September 1904, just a few weeks later, the International Congress of Arts and Science gathered at St. Louis, Missouri, for an international exposition to celebrate progress. Delivered in the midst of a World's Fair built to model the world (here Paris, there London, Turin, New York), Poincaré's speech on the future of physics appropriately aimed to encapsulate the whole field and to identify its weak points. Time coordination featured prominently, set within a larger frame of progressive continuity: "[Y]es, there are indications of a serious crisis [in physics]," Poincaré conceded early on in his talk, but "let us not be too disturbed. We are assured that the patient will not die of this sickness and we can even hope that the crisis will be beneficial, for past history seems to guarantee it."[73]

By Poincaré's lights, mathematical physics had received its first ideal form in Newton's law of gravity. Every body in the Universe, every grain of sand, every star was attracted to every other body by a force inversely proportional to the square of their separation. This simple law, varied and applied to different kinds of forces, formed the first phase of the history of physics. Out of that peaceable kingdom a crisis emerged when Newton's picture proved inadequate to the complex industrial processes physicists confronted in the nineteenth century. New principles were needed, principles that could characterize the whole of a process without specifying, as Newton might have had it, every last detail of the machine. Such new principles included the stipulation that the mass of a system always stayed the same or that the energy of a system remained constant over time. One great triumph was Maxwell's theory, which embraced all of optics and electricity, describing the whole as states of a great, world-pervading ether.

Did this nineteenth-century, second phase of physics far exceed Newton's dreams? Of course. Did it show the futility of the first phase? Poincaré advised his World's Fair audience, "Not in the least. Do you think that this second phase could have existed without the first?" Newton's idea of central forces had led to the principles of the second period (Poincaré's own). "It is the mathematical physics of our fathers that familiarized us little by little with these various principles, which has habituated us to recognize them under the different clothing with which they disguise themselves."[74] Our forebears compared the principles with experience, learned how to modify their expression to adapt them to the givens of experience, enlarged the principles, and consolidated them. Eventually we came to see these principles, including the conservation of energy, as experimental truths. Bit by bit the older conception of central force came to seem superfluous, even hypothetical. The frame of Newtonian physics shook.

Poincaré's staging of this new "crisis" was, characteristically for him, a preamble to recuperation. Echoing the meliorist position he had taken for fifteen years, he insisted that shedding past beliefs did not demand rupture: "The frames are not broken, because they are elastic; but they have been enlarged; our fathers, who established them, have not worked in vain; and we recognize in the science of today the general traits of the sketch that they traced."[75] For Poincaré, physics in 1904 nonetheless stood at the threshold of a new phase in its epochal history. Radium, "that great revolutionary," had destabilized accepted truths of physics, precipitating crisis. Every principle of nineteenth century physics tottered.

One hope had been that meaning could be given to the ether by measuring how fast the earth was sailing through it. Poincaré lamented that all such attempts, even Michelson's most accurate ones, had ended in failure, driving theorists to the limit of their ingenuity. "[I]f Lorentz has managed to succeed, it is only by accumulating hypotheses." Of all those Lorentzian "hypotheses," one stood out for Poincaré above all others:

> [Lorentz's] most ingenious idea was that of local time. Let us imagine two observers who want to set their watches by optical signals; they exchange their signals, but as they know that the transmission is not instantaneous, they take care to cross them. When the station B receives the signal of station A, its clock must not mark the same time as station A at the moment of the emission of the signal, but rather that time augmented by a constant representing the duration of the signal.

At first, Poincaré considered the two clock-minders at A and B to be at rest—their two observing stations were fixed with respect to the ether. But then, as he had since 1900, Poincaré proceeded to ask what happened when the observers are in a frame of reference moving through the ether. In that case "the duration of the transmission will not be the same in the two directions, because station A, for example, moves towards any optical perturbation sent by B, while the station B retreats from a perturbation by A. Their watches set in this manner will not mark therefore true time, they will mark what one can call *local time*, in such a way that one of them will be offset with respect to the other. That is of little importance, because we don't have any way to perceive it." True and local time differ. But nothing, Poincaré insisted, would allow A to realize that his clock will be set back relative to B's, because B's will be offset by precisely the same amount. "All the phenomena that will be produced at A for example, will be set back in time, but they will all be set back by the same amount, and the observer will not be able to perceive it because his watch will be set back; thus, as the *principle of relativity* would have it, there is no means of knowing if he is at rest or in absolute movement."[76]

Yet coordinated clocks were not enough by themselves to rescue all of classical, principle-based physics. According to Poincaré, the challenges of radioactivity hovered like storm clouds over the discipline. Energy conservation: challenged by the spontaneous emission of energetic radioactive particles. Mass conservation: in trouble because fast charged particles acted as if their mass depended on their velocity. Action-reaction: threatened because, according to the Lorentz theory, a lamp projecting a beam of light recoiled *before* the light beam arrived somewhere else where it caused the absorber to recoil. Even the relativity principle appeared threatened, for which the physicist-physicians had called in the antidotes of "local time" and length contraction. What to do? Surely, trust the experimenters. Yet for Poincaré the chain of responsibility ended with the theorists: theorists had produced this mess; they should resolve it. And any theoretical rescue of principled physics could not take place by abandoning the "physics of our fathers." Instead, progress demanded reworking the past: "let us take the theory of Lorentz, turn it in every direction; modify it little by little, and perhaps everything will work out." Poincaré hoped that the organism of physics would reveal its constant identity even as it changed, like an animal shedding its shell for a new one. To abandon a principle such as the principle of relativity would be, he insisted, to sacrifice "a precious arm" in the battle at hand.[77]

Poincaré and Lorentz did "turn the theory in every direction," altering it as best they could. In May 1904, Lorentz modified his old assumption about length contraction and his fictional "local time" in such a way that, when the shortened length and local time were inserted into the equations of physics, the equations were no longer *approximately* the same in any frame of reference moving inertially through the ether, but instead *identical*.[78] That striking vindication of Poincaré's understanding of the relativity principle—as he called it—was enough to set the French polymath to work. On 5 June 1905, he summarized his results to the French Academy of Sciences. For the first time, he had a theory adequate to account for both optical experiments and the new fast electron experiments, both, as he had long hoped, inside the "elastic frame" of Lorentz's physics. Published fully in 1906 as "On the Dynamics of the Electron," the paper finished Poincaré's longstanding project by pushing the clock synchronization scheme one last step: The synchronization of clocks led to Lorentz's improved local time, from which it followed that the equations of physics took on the same form *in all frames of reference.*

Just a few months later, in the winter semester of 1906–07, Poincaré spelled out for his students precisely how Lorentz's improved "local" time fit with the Lorentz contraction to make it fully impossible to detect motion of the earth with respect to the ether.[79] Again in 1908 he insisted that the *apparent* time of transmission is proportional to the *apparent* distance: "it is impossible to escape the impression that the principle of relativity is a general law of Nature, that one could never by any imaginable means, have evidence of anything but the *relative* speeds of objects"—motion with respect to the ether would never be found.[80] Here was a monument to Poincaré's decades-long attempt to improve the machinery of physics while keeping the "elastic frame" of the old—a "new mechanics" that guarded the ether while challenging old ideas of space, time, simultaneity. It was the modernism of an abstract machine.

However beautiful the theory in its embodiment of the relativity principle, Poincaré had long made it clear that principles rose from experiment, acknowledging that those roots meant that experiments could spell trouble for principles. The relativity principle was no exception. Indeed, at the very beginning of his "On the Dynamics of the Electron," Poincaré warned that the entire theory could be endangered by new data.[81] Lorentz too smelled trouble from the laboratory. In a letter to Poincaré of 8 March 1906, Lorentz took pleasure in the coincidence of their results. But that concordance meant that both of them faced a similar peril: "Unfortunately, my hypothesis of the flattening of electrons is in contradiction with the results of new experiments by Mr. [Walter] Kaufmann and I believe I am obliged to abandon it; I am therefore at the end of my Latin and it seems to me impossible to establish a theory that demands a complete absence of the influence of translation on electromagnetic and optical phenomena."[82]

By the time Poincaré and Lorentz put these worries on paper, both saw the essentials of their physics in mortal danger—their hunt for an explanation of the microphysics of the electron, their long struggle to instantiate the relativity principle, and their never-ending quest to specify the status of the ether. In the early years after 1905, some physicists assimilated the new theory to an older ambition to find an electrical explanation for all physics. Others seized the mathematics, ignoring the reformation of time. But in the end, the greatest threat did not emerge from the magnets, tubes, and photographic plates of Kaufmann's study of how fast electrons were deflected by electric and magnetic fields or work at any other laboratory. Though the "true relations" of his physics survived, in the long term Poincaré's vision of the time and space of modern physics—one that prized the ether as an ineliminable framework for intuitive mathematized understanding, one that sharply split true from apparent time—lost its place in the canonical presentation of physics. An unknown twenty-six-year-old physicist at the Bern patent office was to offer a different path to understanding, one that threw aside electron structure, the ether, and the distinction between "apparent" and "true time." Instead he wired a coordinated clock into the theory not as an aid to the physical interpretation of local time but as the capstone of the relativistic arch.

Chapter 5

EINSTEIN'S CLOCKS

Materializing Time

I N JUNE OF 1905, the contrast between Einstein and Poincaré could not have been greater. Poincaré was an Academician in Paris, fifty-one years old and at the height of his powers. He had been a professor at France's most illustrious institutions, run inter-ministerial commissions, and published a shelf of books—volumes on celestial mechanics, electricity and magnetism, wireless telegraphy, and thermodynamics. With over 200 technical articles to his name, he had altered whole fields of science. His best-selling volume of philosophical essays had brought his abstract reflections on the meaning of science to a huge audience, including Einstein. Einstein, at twenty-six, was by contrast an unknown patent officer, living in a walk-up flat in a modest section of Bern.

Unlike France or Britain, Switzerland was not a colonial power. Unlike the United States or Russia, it had no vast longitudinal spread, nor did it have even a hectare of unsettled land to be colonized by railways, telegraphs, and time links. In fact, Switzerland was late to adopt the telegraph and, relative to other European countries, late in the construction of a rail network. But when rails and telegraphs did come to this mountainous country in the latter part of the nineteenth century, the movement to synchronize clocks quickly gained momentum—not surprising in a nation in which the production of precise timepieces was, by century's end, an urgent matter both of national pride and economic significance.[1]

The clock-world famous Matthäus Hipp found welcome in Switzerland. Blacklisted in his native Württemberg for his republican and democratic advocacy around 1848, his trade was in time machines of every sort. He developed electrically maintained pendula so regular that they far surpassed mechanical ones: a heavy pendulum swung freely except when it needed an electromagnetic boost. Alongside such electrical timepieces, Hipp perfected recording clocks that radically altered experimental psychology. Collaborating with physicists and astronomers, he tracked the transmission speeds of nerves, telegraphs, and light, thereby inventing, altering, and producing new ways to use electricity and clocks to materialize time. Though he worked closely with scientists (especially Swiss astronomer Adolphe Hirsch), Hipp was more artisan-entrepreneur than mathematician-savant. Founder of telegraph and electrical apparatus factories in Bern and later Neuchâtel and Zurich, Hipp took his company from the establishment of the first network of public electric clocks in Geneva (1861) to ever greater prominence. In 1889 Hipp's firm became A. De Peyer and A. Favarger et Cie; from then to 1908 the concern extended the range of their mother clocks beyond the dominion of observatories and railways to steeple clocks and even to the wake-up clocks inside hotels.[2] With the march of time into every street, engineers needed methods to extend indefinitely the number of units that could be branched together. A flood of patents followed, perfecting relays and signal amplifiers.

If you wanted to display time on a major building, you needed more than one clock before time unification. The Tower of the Island in Geneva boasted three around 1880: a big clockface in the center showed middle Geneva time (about 10:13); the face on the left showed Paris time for the Paris-based "Paris-Lyon-Mediterranean" line (9:58); and the right-hand clock boasted Bern time, a handsome five minutes in advance of Geneva (10:18). Clock synchronization in Switzerland was public and eminently visible. So, too, was the chaos of uncoordinated time.

Bern inaugurated its own urban time network in 1890; improve-

Figure 5.1 *Three Clocks: Tower of the Island of Geneva (circa 1880). Before time unification, a fine clock tower like this one publicly registered the multiplicity of times.* SOURCE: CENTRE D'ICONOGRAPHIE GENEVOISE, RVG N13X18 14934.

ments, expansions, and new networks sprouted throughout Switzerland. Not only was accurate, coordinated time important for European passenger railroads and the Prussian military, it was equally crucial for the dispersed Swiss clockmaking industry, which desperately needed the means for consistent calibration.[3] But time was always practical and more than practical, material-economic necessity *and* cultural imaginary. Professor Wilhelm Förster of the Berlin Observatory, which set the Berlin master clock by the heavens, sniffed that any urban clock that did not guarantee time to the nearest minute was a machine "downright contemptuous of people."[4]

Einstein's patent-office window on this electrochronometric world opened a crucial moment in the synchronization of Swiss

Figure 5.2 One Clock: Tower of the Island of Geneva (after 1886). After time unification, the same tower shown in figure 5.1 needed but a single clock: time unification was visible for all to see. Source: Centre d'Iconographie Genevoise, RVG N13x18 1769.

time. For despite General von Moltke's resounding support for a pan-German time unification and the undamped enthusiasm of the North American advocates of one-world time, Albert Favarger, one of Hipp's chief engineers and the man who effectively succeeded Hipp at the helm of his company, was not at all satisfied with the rate of progress. He intended to say so, very publicly, at the 1900 Exposition Universelle in Paris. Here, the International Congress on Chronometry met to discuss the status, inter alia, of clock-coordination efforts.[5] At the outset of his speech to the congress, Favarger asked how it could be that the distribution of electrical time was running so distressingly far behind the related technologies of telegraphy and telephony? First, he suggested, there were technical difficulties; remotely coordinated clocks could rely on no obliging friend ("ami complaisant") to oversee and correct the least difficulty, whereas the steam engine, dynamo, or telegraph all seemed to run with constant human companionship. Second, there was a technician gap—the best technical people were staffing power and communication devices, not time machines. Finally, he lamented, the public was not funding time distribution as it should. Such lagging boosterism baffled Favarger: "Could it be possible that we have not experienced the imperious, absolute, I would say collective need of time exactly uniformly, and regularly distributed? . . . Here's a question that borders on impertinence when addressed to a late 19th century public, laden with business and always rushed, a public that has made its own the famous adage: Time is money."[6]

As far as Favarger was concerned, the sorry state of time distribution was out of all proportion with the exigencies of modern life. He insisted that humans needed exactitude and universality correct to the nearest second. No old-fashioned mechanical, hydraulic, or pneumatic system would do. Electricity was the key to the future, a future that would only come about properly if humankind broke with its past of mechanical clocks: a technical era riven by anarchy, incoherence, and routinization. In place of the pneumatic chaos of Paris or Vienna, the new world of electrocoordinated clocks would be based on a rational and methodical approach. As he put it,

You don't have to run long errands through Paris to notice numerous clocks, both public and private, that disagree—which one is the biggest liar? In fact if even just one is lying one suspects the sincerity of them all. The public will only gain security when every single clock indicates unanimity at the same time at the same instant.[7]

How could it be otherwise? In terms reminiscent of time struggles in the United States some years earlier, Favarger reminded the assembled exposition attendees that the speed of trains roaring through Europe was mounting to 100, 150, even 200 kilometers per hour. Those running the trains and directing their movements, not to speak of the passengers entrusting their lives to such speeding carriages, had to have correct times. At 55 meters per second, every tick counted, and the prevalent but obsolete mechanical systems of coordination were bound to be inferior. Only the electric, automatic system was truly appropriate: "The nonautomatic system, the most primitive yet the most widespread, is the direct cause of the time anarchy that we must escape."[8]

Time anarchy. No doubt Favarger's reference recalled in his listeners the anarchism that had taken a powerful hold among the Jura watchmakers (or for that matter it might have reminded Parisian listeners of French anarchist Martial Bourdin's detonation outside Greenwich Observatory). Peter Kropotkin had broadly publicized the clockmakers' anarchism just a year before in his 1898–99 *Memoirs of a Revolutionist*:

> The equalitarian relations which I found in the Jura Mountains, the independence of thought and expression which I saw developing in the workers, and their unlimited devotion to the cause appealed even more strongly to my feelings; and when I came away from the mountains, after a week's stay with the watchmakers, my views upon socialism were settled. I was an anarchist.[9]

Favarger was, however, more concerned about an anarchism signaled by a broader disintegration of personal and societal regularity. Not for him the older pneumatic system—those steam-driven, rigidly framed branched pipes that had pulsed compressed-air time to fixed public clocks and private timepieces in Vienna and Paris. Only electrical distribution of simultaneity could provide the "indefinite expansion of the time unification zone."[10] Favarger's unwavering support for distant simultaneity issued from many sources, from the practicalities of train scheduling and the entrepreneurial ambitions of his company to a sense of what time would mean for the inner life of the modern citizen. Time synchronization was all at once political, profitable, and pragmatic.

Should we escape this dreaded anarcho-clockism, Favarger assured his listeners, we could fill a great lacuna in our knowledge of the world. For even as the Paris-based International Bureau of Weights and Measures had begun to conquer the first two fundamental quantities — space and mass — Favarger insisted that the final frontier, time, remained unexplored.[11] And the way to conquer time was to create an ever-widening electrical network, bound to an observatory-linked mother clock that would drive relays multiplying its signals and send automatic clock resets into hotels, streetcorners, and steeples across the continents. Affiliated with Favarger was a company determined to synchronize Bern's own network. When, on 1 August 1890, Bern set the hands of its coordinated clocks in motion, the press hailed the "Revolution in Clocks."[12]

Even today, from many places in Bern you can clearly see the faces of several grand public clocks. In that August of 1890, when they all began running in step, the order of coordinated time was written upon the architecture of this city of arcades and churches. Swiss clockmakers publicly joined the worldwide project of electric simultaneity.

Theory-Machines

During the 1890s Einstein was not yet concerned about clocks at all; but as a young man of sixteen, in 1895, he was very much concerned with the nature of electromagnetic radiation. That summer Einstein put on paper his reflections of how the state of the ether would alter in the presence of a magnetic field; for example, how its parts would distend in response to a passing wave. Even his untutored imagination was uncomfortable with the "customary conception" of radiation as a wave in static, substantial ether. Suppose, he later recalled thinking, that one could catch up to a light wave and ride it, so to speak, as classical physics might imply. Then one would see the electromagnetic wave unfold before him, the field undulating in space but frozen in time. But nothing like such a frozen wave had ever been observed.[13] Something was wrong with this way of thinking, but Einstein did not know what.

After a first unsuccessful application, Einstein began his training at Switzerland's (and one of Europe's) great technical universities: the Eidgenössische Technische Hochschule (ETH), founded in 1855. Certainly the ETH of 1896 was a very different place from the Ecole Polytechnique that Poincaré had entered in the early 1870s. To be sure, both stressed engineering. But Ecole Polytechnique's fame had long rested on its schooling the elite in a concentrated mix of pure mathematics and scientific training, a foundation on which its graduates would then build at places like the School of Mines. For the French ever since Napoleon, the ambition had been to educate an elite in high mathematics that would be (in due time) able to meet the demands of the practical world they would control. Founded in the mid-nineteenth century in a Switzerland short on natural resources and long on desire to catch up with the rapid industrialization of France, Britain, and Germany, ETH was very different. ETH wanted an immediate link between theory and praxis. There was never a moment when the demands of road, railroad, water, electrical, and bridge construction were lost to sight or even put in the background.[14] Take mechanics. At Ecole Polytechnique, Poincaré celebrated the subject as the "factory stamp" for all students, from those striving to become abstract mathematicians to those whose ambitions would take them into administrative or military service. Mathematics was the queen of the university, structuring the teaching of mechanics so that its abstractions were only gradually specified to the point where it met application. By contrast, in Zurich, it was a mining engineer with little interest in abstract mathematics who had set the tone of the mechanics course back in 1855. Built more along German pedagogical lines than along French ones, the Swiss insisted that the abstract should not wait for an eventual union with the applied in a next stage of education. Swiss industrialists needed help building everything from telegraphs and trains to water works and bridges. From the start (and throughout its history) the applied and abstract entered ETH together.

So, while Poincaré and his contemporaries learned about experimentation by watching demonstrations at the front of the amphitheater, Einstein spent a great deal of time learning hands-on in the well-appointed physics laboratory at ETH. Formal treatments of the principles of devices were typical of the work at Polytechnique; when Cornu wanted to study synchronized clocks, he wrote down an elegant theory of the physics that underlay them. Instead, when Einstein's physics teacher Heinrich Friedrich Weber taught, he spoke about the precise heat-conducting ability of granite, sandstone, and glass. Thermodynamics at ETH oscillated back and forth between the basic equations and detailed numerical calculations and laboratory arrangements of glass, pumps, and thermometers, as Einstein carefully recorded in his notebooks.[15] Indeed, differences between the two institutions reflected their views about what theories said about the world. In Poincaré's Ecole Polytechnique, he would have found a proud agnosticism toward atoms (or many other hypothetical physical objects). At ETH, Weber and his colleagues had no time for such fancy dancing, no interest in exploring, for its own sake, the myriad ways one could account for a collection of phenomena. After introducing "heat" without concern for its "true nature," Weber argued that the connections among physical quantities led directly to a mechanical picture in which heat was nothing more than molecular motion. Then he calculated numbers of molecules and fixed their properties. No metaphysical realism, just a matter-of-fact engineer's assessment that atoms let the work proceed.[16]

In the summer of 1899, Einstein still was agonizing over the ether, moving bodies, and electrodynamics. To his beloved Mileva Marić, he recalled that back in Aarau (in secondary school), he had thought up a way to measure, and perhaps to explain, how light traveled in transparent bodies when these transparent bodies were dragged through the ether.[17] Now he conveyed to her his sense that naive, material ether theories, ones that proposed bits of ether here and there moving this way and that, would simply have to go. No doubt he had absorbed some of this austere attitude toward theory through the emphasis on measurement at ETH. Still, the school clearly frustrated Einstein by not offering more about the relatively new Maxwellian theories of electricity and magnetism. So he began teaching himself, and one clearly important source was the work of Heinrich Hertz. Hertz had stripped Maxwell's complicated theory of electricity and magnetism down to its bare-boned equations and, to general astonishment, demonstrated experimentally the existence of electric (radio) waves in the ether. Throughout his short life, Hertz paid extraordinary attention to the different ways of formulating theories of electricity and magnetism, doubting out loud that the "name" electricity or the "name" magnetism corresponded to anything substantial in its own right. Einstein then turned Hertz's critical sword against the still-vibrant ether:

I returned the Helmholtz volume and am now rereading Hertz's propagation of electric force with great care because I didn't understand Helmholtz's treatise on the principle of least action in electrodynamics. I'm convinced more and more that the electrodynamics of moving bodies as it is presented today doesn't correspond to reality, and that it will be possible to present it in a simpler way. The introduction of the name "ether" into theories of electricity has led to the conception of a medium whose motion can be described, without, I believe, being able to ascribe physical meaning to it.[18]

Electricity, magnetism, and currents, Einstein concluded, should be definable not as alterations of a material ether but as the motion of "true" electrical masses with physical reality through empty space. A motionless, nonmaterial conception of the ether might work better than a material one, and (following Lorentz's widely hailed theory) many leading physicists had just such a conception in mind.

Faced with the inability of experimentalists to either drag the ether or detect motion through it, Einstein sometime around 1901 disposed even of this static insubstantiality. Ether, that centerpiece of nineteenth-century physical theories, was gone. For Einstein it was neither the ultimate constituent of electrical particles nor the all-pervasive medium necessary for light to propagate. Even before Einstein had set foot through the patent office door, crucial pieces of the puzzle were in place; he may well have already begun invoking the relativity principle.[19] Certainly he was reconsidering the meaning of Maxwell's equations while latching onto a realistic picture of moving electrical charges. He had dismissed the ether. Still, none of these considerations directly bore on *time*.

From 1900 through 1902, Einstein struggled at the margins of institutionalized science. Despite having received his mathematics-teaching diploma from ETH in July 1900, he found himself unable to secure a university job. He took up tutoring and began pushing, outside university walls, into two domains of theoretical physics. On the one hand, he explored the nature of thermodynamics (the science of heat) along with its foundation and extension through statistical mechanics (the theory that heat was nothing but the motion of particles). On the other hand, he strove, not yet in print, to grasp the nature of light and its interaction with matter. Above all else, he wanted to know how electrodynamics would look for moving bodies.

Einstein's relentless optimism and self-confidence, combined with a biting disregard for complacent scientific authority, shows in a myriad of letters. In May 1901, he confided to Mileva that "Unfortunately, no one here at the Technikum [the ETH] is up to date in modern physics & I have already tapped all of them without success. Would I too become so lazy intellectually if I were ever doing well? I don't think so, but the danger seems to be great indeed."[20] Or the next month, having written up a custom-made critique of Paul Drude (a leading figure in the theory of electrical conduction), Einstein told Mileva: "What do you think is lying on the table in front of me? A long letter addressed to Drude with two objections to his electron theory. He will hardly have anything sensible to refute me with, because the things are very simple. I am terribly curious whether and what he is going to reply. Of course, I also let him know that I don't have a job, that goes without saying." If Einstein wouldn't budge an inch on physics, he had no more intention of altering his personal life to suit the disapproving glances of his nearest and dearest. When friends apparently criticized his personal comportment, he rejected their judgment out of hand: "imagine the Wintelers railed against me . . . & said that I have been leading a life of debauchery in Zurich."[21]

From early in his time at Polytechnique, Poincaré had formed lifelong bonds with his teachers. He admired his elders, filially naming many of his own mathematical creations after them. Einstein, by contrast, appears to have been absolutely undeterred by the head-shaking of his old teachers, his incessant job rejections, or, for that matter, his mother's stinging disapproval of Mileva Marić. So when Drude dismissed Einstein's objections in July 1901, the younger scientist simply relegated him to the throng of clay-footed authority:

There is no exaggeration in what you said about the German professors. I have got to know another sad specimen of this kind—one
of the foremost physicists of Germany. To two pertinent objections
which I raised against one of his theories and which demonstrate
a direct defect in his conclusions, he responds by pointing out that
another (infallible) colleague of his shares his opinion. I'll soon
make it hot for the man with a masterly publication. Authority
gone to one's head is the greatest enemy of truth.[22]

Einstein might "make it hot"; but these authorities were not about
to respond to his thermo-pressure with a shower of job offers. One
after another, rejections arrived, including one for the position of
senior teacher, Mechanical Technical Department in the Cantonal
Technikum at Burgdorf.[23] When his friend the mathematician Marcel Grossmann landed a position at the Cantonal School in Frauen-
feld, Einstein congratulated him heartily, adding that the security
and good work it afforded would certainly be welcome. Einstein,
too, had tried applying. "I did it only so that I wouldn't have to tell
myself that I was too faint-hearted to apply: for I was strongly convinced that I have no prospects of getting this or another similar
post."[24]

Then came a genuine prospect of employment. The Swiss
Patent Office in Bern had placed an advertisement for an opening.
Einstein wrote directly: "I, the undersigned, take the liberty of
applying for the position of Engineer Class II at the Federal Office
for Intellectual Property, which was advertised in the Bundesblatt
[Federal Gazette] of 11 December 1901."[25] Assuring the patent
office that he had studied physics and electrical engineering at the
School for Specialist Teachers of Mathematics and Physics at ETH,
he promised that all the documents were ready and waiting. On 19
December 1901, he rejoiced to Mileva: "But now listen & let me
kiss you and hug you with joy! [Friedrich] Haller [head of the
patent office] has written me a friendly letter in his own hand, in

which he requested that I apply for a newly created position in the patent office! Now there is no longer any doubt about it. [Former ETH classmate, Marcel] Grossmann has already congratulated me. I am dedicating my doctoral thesis to him, to somehow express my gratitude."[26] Job in hand, or almost so, he and Mileva could marry.

Even his attitude toward his old thesis evaluator, Kleiner, brightened. On 17 December he had told Mileva that he'd "descend upon . . . that bore Kleiner." Einstein wanted permission to work during Christmas vacation. "To think of all the obstacles that these old philistines put in the way of a person who is not of their ilk, it's really ghastly! This sort instinctively considers every intelligent young person as a danger to his frail dignity, this is how it seems to me by now. But if he has the gall to reject my doctoral thesis, then I'll publish his rejection in cold print together with the thesis & he will have made a fool of himself. But if he accepts it, then we'll see what a position the fine Mr. Drude will take . . . a fine bunch, all of them. If Diogenes were to live today, he would look with his lantern for an *honest* person in vain."[27]

Two days later, Diogenes's little lantern cast humanity in a warmer light: "Today I spent the whole afternoon with Kleiner in Zurich and otherwise talked with him about all kinds of physical problems. He is not quite as stupid as I thought, and moreover, he is a good guy." True, Kleiner had not yet read his thesis, but Einstein was not worried. His attention had already shifted elsewhere: "[Kleiner] advised me to publish my ideas about the electromagnetic theory of light in moving bodies together with the experimental method. He thought that the experimental method proposed by me is the simplest and most appropriate one conceivable."[28]

Encouraged, no doubt, by this unexpected support, Einstein dug deeper into the theories of the ether and motion within it. He resolved to study Lorentz and Drude on the electrodynamics of moving bodies (it is perhaps a measure of his disconnection from the mainstream that he had not already done so) and borrowed a physics volume from his friend Michele Besso. Poincaré endlessly stressed in his lectures of 1899 how valuable older physics approaches were. Maxwell, Hertz, Lorentz, and many others were all worth studying in detail because even when parts misfired, each captured "true relations" among physical quantities. That form of patience was not for Einstein. One ether-theory text impressed him only by its obsolescence, as he told Mileva late in 1901: "Michele gave me a book on the theory of the ether, written in 1885. One would think it came from antiquity, its views are so obsolete. It makes one see how fast knowledge develops nowadays."[29] Not long before, he had told her that he and Besso had been pondering together "the definition of absolute rest."[30]

In the early weeks of 1902, Einstein moved his modest collection of household items from Schaffhausen (where he had a temporary post in a private school) to Bern, and began once again to search for tutorial students on 5 February 1902:

> Private lessons in
> MATHEMATICS AND PHYSICS
> for students and pupils
> given most thoroughly by
> ALBERT EINSTEIN, holder of the fed.
> Polyt. Teacher's diploma
> GERECHTIGKEITSGASSE 32, 1ST FLOOR
> Trial lessons free.[31]

A few days later his prospects looked good. The young tutor's notice netted two students, Einstein was planning on writing to the great expert on statistical mechanics, Ludwig Boltzmann, and his studies outside of a narrow band of physics were proceeding apace: "I have almost finished reading [Ernst] Mach's book with tremendous interest, and will this evening."[32]

Maurice Solovine, who had come from Romania to study at the University of Bern, was one new student, along with another friend of Einstein's, Conrad Habicht, who was pursuing his doctorate in mathematics. Together the three founded their "Akademie Olympia," an informal club for discussing philosophy or whatever else aroused their curiosity. Solovine: "We would read a page or a half a page—sometimes only a sentence—and the discussion would continue for several days when the problem was important. I often met Einstein at noon as he left his desk and renewed the discussion of the previous evening: 'You said . . . , but don't you think . . . ?' Or: 'I'd like to add to what I said yesterday. . . .'"[33]

Mach was on the agenda. Mach the philosopher-physicist-psychologist stood for a relentlessly critical attitude toward that which could not be made accessible to the senses. Though Einstein never subscribed entirely to what he considered Mach's overemphasis on the sensory, Einstein did pull from Mach's writings a critical club to be wielded against idle metaphysical chatter such as the notions of "absolute time" and "absolute space."[34] In Einstein's favorite work of Mach's (*The Science of Mechanics*, 1883) he would have found a polemic against Newton's absolute time and space that began by citing the master: "Absolute, true, and mathematical time, of itself, and by its own nature, flows uniformly on, without regard to anything external. . . . Relative, apparent, and common time is some sensible and external measure of absolute time." Such thoughts disclosed, according to Mach, not the Newton of hard-boiled facts, but the Newton of medieval philosophy. For Mach, time was not something primitive against which phenomena were measured. Quite the opposite: time itself derived from the motion of things—the earth as it spins, the pendulum as it swings. To attempt to get behind the phenomena to the absolute was futile. Mach's condemnation was clear: "This absolute time can be measured by comparison with no motion; it has therefore neither a practical nor a scientific value; and no one is justified in saying that he knows aught about it. It is an idle metaphysical conception."[35]

Over the following years, Einstein often emphasized how important this time analysis of Mach was for him. For example, in a 1916 article memorializing Mach, Einstein insisted: "These quotations [from *The Science of Mechanics*] show that Mach clearly recognized the weak points of classical mechanics, and thus came close to demanding a general theory of relativity—and this almost half a century ago! It is not improbable that Mach would have hit on relativity theory if in his time . . . physicists had been stirred by the question of the meaning of the constancy of the speed of light. In the absence of this stimulation, which flows from Maxwell-Lorentzian electrodynamics, even Mach's critical urge did not suffice to raise a feeling for the need of a definition of simultaneity for spatially distant events." As it had for Poincaré, simultaneity for Einstein crossed electrodynamics with philosophy.[36]

Up for discussion by the Olympia Academy was also Karl Pearson, the Victorian mathematician-physicist known for his contributions to statistics, philosophy, and biology. But intriguingly Pearson, drawing on Mach as well as the German philosophical tradition, also put the naive reading of "absolute time" under the critical microscope. Einstein and his academicians would have found in Pearson's 1892 *Grammar of Science* another sharply critical assessment of the relation of all observable clocks to Newton's absolute time: "[T]he hours on the Greenwich astronomical clock, and ultimately on all ordinary watches and clocks regulated by it, will correspond to the earth turning through equal angles on its axis." So all clock time is ultimately astronomical time. But many factors could alter the great earth-clock in its rotation—tides, for example. "Absolute, true, mathematical time," as Newton called it, is something we use to describe our sense impressions (a "frame," as Pearson put it). "But in the world of sense-impression itself [absolute intervals of time] have no existence." Watching stars cross the spider lines of our transit instruments from one midnight to the next on two different occasions, we notice that the average person registers roughly the same sequence of sense impressions. Fine, says Pearson. Assign those two midnight-to-midnight intervals the designation "equal." But we should not fool ourselves; this has nothing to do with absolute time: "The blank divisions at the top of and bottom of our conceptual time-log are no justification for rhapsodies on the past or future eternities of time, for rhapsodies which, confusing conception and perception, claim for these eternities a real meaning in the world of phenomena, in the field of sense impression."[37]

Along with Mach and Pearson, Richard Avenarius's *Critique of Pure Experience* found its way to the Olympians' reading list; it, too, took a skeptical attitude toward that which lay outside the grasp of experience. Other works on the discussion roster included John Stuart Mill's *Logic*, which cautioned against "the prevailing hypothesis of a luminiferous ether"; Dedekind's sharp analysis of the number concept; and David Hume's dismantling of induction in the *Treatise on Human Nature*.[38] But of all these works that framed the basic notions of science against a critical background of that to which the human mind had access, Solovine singled out one book for special commentary, a volume that had just appeared in German in 1904: ". . . Poincaré's *Science and Hypothesis*, which engrossed us and held us spellbound for weeks."[39] From it Einstein and his Olympians could find strong backing for Mach's and Pearson's views. Poincaré also alludes to his crucial article, "The Measure of Time."

It may be that Einstein and his academy raced off to find "The Measure of Time" in its original form (published in a French philosophy journal). That seems unlikely. But there's an intriguing twist. Unlike the English or French versions, the German publishers of *Science and Hypothesis* translated and included a good-sized excerpt from the conclusion to "Measure of Time" as a footnote. So in 1904, in plain German, the Olympia Academy would have had before them Poincaré's explicit denial of any "direct intuition" about simultaneity, his insistence that rules defining simultaneity were chosen for convenience not truth, and his final pronouncement: "All these rules and definitions are only the fruit of an unconscious opportunism." In fact, the French-to-German translators went further, providing references to the long line of philosophers, physicists, and mathematicians who had blasted absolute time in favor of relative time, with Locke and d'Alembert leading the way and with one of the creators of non-Euclidean geometry, Lobaschewski, there to report that time was just "motion designed to measure other motions." Pendulum clock, spring clock, spinning earth: Poincaré's German translators insisted that we had our choice of which motion to use as the unit of time that would define the time "t" of physics. But that choice had nothing whatsoever to do with absolute time. Instead, the choice underscored yet again Poincaré's argument for "opportunism" in the physical definition of simultaneity and duration.[40]

While it would be a mistake to assume that the finer details of any single philosopher's thought leapt whole into Einstein's work, there is no doubt at all that he emerged from these early Bern years with a powerful sense of the distinction between that which was accessible to our experience and that which was, so to speak, inaccessibly hidden behind the curtain of the perceptible. Such an emphasis on the knowable, especially as knowable through that which can be grasped of the natural world through the senses, was key to the doctrine of positivism laid out by Auguste Comte and pursued by his many followers. As Einstein told philosopher Moritz Schlick in 1917: "Your representations that the theory of rel[ativity] suggests itself in positivism, yet without requiring it, are . . . very right. In this also you saw correctly that this line of thought had a great influence on my efforts, and more specifically, E. Mach, and even more so Hume, whose *Treatise of Human Nature* I had studied avidly and with admiration shortly before discovering the theory of relativity. It is very possible that without these philosophical studies I would not have arrived at the solution."[41]

Philosophy mattered. Far from ungrounded in a cultural Zeitgeist, Einstein's advocacy for a procedural notion of simultaneity and against metaphysical, absolute time fit directly into this series of moves to shore up the basis of physical knowledge by Mach, Pearson, Mill, and Poincaré. One by one Einstein and his little circle sought out their books for discussion.

Einstein's first important physics papers (1902–04) targeted thermodynamics and the underlying account of heat as the product of molecular motion (statistical mechanics). Somewhere among his explorations of philosophy, his immersion in thermodynamics at ETH, and his independent post-ETH research, Einstein began to forge an approach to physics that emphasized principles and eschewed detailed model building. Famously, thermodynamics could be formulated with two easily stated and yet grand claims about isolated systems: the amount of energy always remains the same, and entropy (the disorder of a system) never decreases. The simplicity and scope of this theory remained for Einstein an ideal of science throughout his career. In Poincaré's *Science and Hypothesis* he would have found a powerful exposition of the view that physics was concerned with the analysis of such principles.

And yet the terms "conventions" and "principles" did not resonate the same way for Einstein (and indeed for many German physicists at the time) as they did for Poincaré. In French, the triple meaning of the French word *convention* (legal accord, scientific agreement, *Convention* of Year II) was vividly present for Poincaré and his circle around the *Convention du Mètre* or the proposed *convention* for decimalized time. Moving the text into German meant breaking that particular chain of associations. Indeed, the German translators of *Science and Hypothesis* split their rendition of *convention*, sometimes offering a German noun that included a legal accord (*Übereinkommen*), sometimes a socially conventional agreement (*konventionelle Festsetzung*).[42] More importantly, we would look in vain in Einstein's works for a Poincaré-like insistence that principles were definitions, surviving purely because of their "convenience." For Einstein principles supported physics and perhaps more than science, especially in later years. In a different context Einstein reflected:

> My interest in science was always essentially limited to the study of principles, which best explains my conduct in its entirety. That I have published so little is attributable to the same circumstance, for the burning desire to grasp principles has caused me to spend most of my time on fruitless endeavors.[43]

Principled physics was important to Einstein. So was the critical philosophical reflection that pared physics concepts to their basic elements. But there was more. Throughout his Bern years, Einstein's work and thought was saturated by the materialized principles of *machines*. From the outset, Einstein regarded the patent office not as a burden interfering with his "real" work, but instead as a productive pleasure. In mid-February 1902 he recounted how he had come upon a former ETH student who was now working at the patent office. "He finds that it's very boring there—certain people find everything boring—I am sure that I will find it nice and that I will be grateful to Haller as long as I live." Or as Einstein later testified, "Working on the final formulation of technological patents was a veritable blessing for me. It enforced many-sided thinking and also provided important stimuli to physical thought."[44]

What about time? By 1902 Lorentz had long since been experimenting with fictional mathematical variations in the way the time variable *t* would be defined for an object moving in the ether. With Poincaré's and other physicists' further articulation, the notion of a fixed ether had gained ground. As we know, Poincaré had (first ignoring the ether altogether and then for systems explicitly moving through it), *interpreted* simultaneity through light-signal-coordinated clocks. Though his use of the ether changed, Poincaré never wavered from his conviction that the ether was enormously valuable as a tool for thinking, a condition for the application of productive intuition. He *never* equated "apparent time" (time measured in a moving frame) with "true time" (time measured at rest in the ether).[45]

Poincaré, Lorentz, and Abraham were determined to *begin* their theorizing with an analysis of dynamics, the forces that held together, flattened, and joined electrons as they ploughed through the ether. Forces contracted the arms of Michelson and Morley's interferometer; forces kept all the negative charge in an electron from blowing that charged particle to smithereens. Out of such *constructive*, built-up theories of matter they wanted to *deduce* kinematics—the behavior of ordinary matter in the absence of external forces.

Einstein's goal was altogether different: time would not begin with dynamics. In mid-1905, he advanced a new account of time and space that would start the physics of moving bodies with simple physical principles the way thermodynamics began with the conservation of energy and a forbidden decrease of entropy. Lorentz was willing to posit an artificial notion of time (t_{local}) because of its utility in calculating solutions to his equations. Poincaré saw the physical consequences of "local time" for frames of reference in constant motion. Before Einstein, however, neither Poincaré nor Lorentz seized the coordination of clocks as the crucial step that would reconcile some of the great principles of physics. Neither expected that reconceiving time would up-end their conception of the ether, electrons, and moving bodies. Neither expected that the Lorentz contraction itself could be viewed as merely consequent to a redefinition of time. For his part, before May 1905 Einstein acknowledged only the basic features of Lorentz's 1895 physics, and not a single piece of work on electrodynamics from Poincaré. Instead, Einstein was reading philosophical texts about the foundations of science (including Poincaré's), publishing on the molecular-statistical theory of heat, and exploring the electrodynamics of moving bodies. Before stepping into the patent office, he left not the slightest clue that he had any interest in clocks, time, or simultaneity.

Patent Truths

So things stood for Einstein when he arrived at the Bern patent office in June 1902.[46] This site represented (and not just for Einstein) not only a job but also a site of training—a rigorous school for thinking machines. Heading the patent office during Einstein's tenure was Friedrich Haller, a stern taskmaster to his underlings, who directed the young inspector to be critical at every stage of his evaluation of proposals: "When you pick up an application, think that anything the inventor says is wrong." Above all, he was warned, avoid easy credulity: the temptation will be to fall in with "the inventor's way of thinking, and that will prejudice you. You have to remain critically vigilant."[47] To Einstein, already amply inclined to treat what he considered arbitrary authority as obsolete, thick-headed, and lazy, the injunction to indulge his skepticism to the full must have been gratifying. His inclination to test complacent assumptions in the use of gears and wires echoed a similarly icono-clastic attitude in less tangible realms of physics. For in the electrodynamics of moving bodies, Einstein had chosen a problem that had challenged him on and off for some seven years and was, with increasing force, preoccupying the leading physicists of the day.

Einstein's work in electrodynamics in 1902 did not include an inquiry into the nature of time. But he was literally surrounded by burgeoning fascination with electrocoordinated simultaneity. Every day Einstein stepped out of his house, turned left, and made his way to the patent office—a walk that brought him to a workplace which, he told one friend, "I enjoy . . . very much, because it is uncommonly diversified and there is much thinking to be done."[48] Every day he had to walk past the great clock towers presiding over Bern, their time coordinated. Every day he passed the myriad of electric street clocks recently, and proudly, branched to the central telegraph office. Strolling from his street, the Kramgasse, to the patent office, Einstein had to walk under one of the most famous clocks of the city. (See figures 5.3 and 5.4.)

Friedrich Haller was in many ways as much Einstein's teacher at the patent office as Weber or Kleiner had been at ETH. Under his

Figure 5.3 Coordinated Clock Tower: Kramgasse. As he stepped out of his Kramgasse apartment and turned left toward the Bern Patent Office, Einstein viewed one of the great (and by 1905, coordinated) clocks of the city. SOURCE: BÜRGERBIBLIOTHEK BERN, NEG. 10379.

Figure 5.4 Bern's Electrical Clock Network (circa 1905). Coordinated, electrical clocks were a matter of practical import and cultural pride. By 1905, they were a prominent piece of the modern urban landscape throughout the city of Bern. SOURCE: BERN CITY MAP FROM THE HARVARD MAP COLLECTION; CLOCK LOCATIONS SHOWN USING DATA FROM MESSERLI, *GLEICHMÄSSIG* (1995).

aegis, the patent office truly was a school for novel technology, a site that aimed to train a quite specific and disciplined taking-apart of technological proposals. Early on Haller reproached Einstein: "As a physicist you understand nothing about drawings. You have got to learn to grasp technical drawings and specifications before I can make you permanent."[49] In September 1903 Einstein apparently had sufficiently conquered the visual language of the patent world; he received notice that his appointment had been made permanent. Still, Haller was not yet ready to promote him, commenting in an evaluation that Einstein "should wait until he has fully mastered machine technology; judging by his course of studies, he is a physicist." That mastery came as Einstein plunged himself into the critical evaluation of the parade of patents that came before him. Soon he could report to Mileva that he was "getting along with Haller better than ever before. . . . When a patent agent protested against my finding, even citing a German patent office decision in support of his complaint, [Haller] took my side on all points."[50] After three-and-a-half years at the patent bench, Einstein persuaded the authorities that, physics notwithstanding, he had learned a new way of looking through diagrams and specifications to the core of innovative technology. In April 1906, Haller promoted him to technical expert, second class, judging that Einstein "belongs among the most esteemed experts at the office"[51] where, increasingly, patents on electric time swelled the number of applications.

Time technologies spun off patents in every sector of the network: patents on low-voltage generators, patents on electromagnetic receivers with their escapements and armatures, patents on contact interrupters. Fairly typical of the kind of electrochronometric work blossoming in the decade after 1900 was Colonel David Perret's novel receiver that would detect and use a direct-current chronometric signal to drive an oscillating armature. It was issued Swiss patent number 30351 on 12 March 1904. Favarger's own receiver did the opposite: it took an alternating current from the mother clock and turned it into the unidirectional motion of a toothed wheel. The request for this patent—later used widely—was stamped "received" on 25 November 1902 and was issued on 2 May 1905 after a long, but not entirely unusual, evaluation period. There were patents that specified systems for clocks to activate remotely placed alarms, while other submissions targeted the distant electromagnetic regulation of pendula. There were proposals for time to travel over telephone lines—even systems that sent time by wireless. Other patents advanced schemes to monitor railroad departures and arrivals or to display the time in different time zones. Yet others specified how remotely run electrical clocks could be protected from atmospheric electricity or how electromagnetic time signals could be silently received. A cascade of coordinated time.

Some of these patents systemically addressed the problem of distributing simultaneity. Perret's patent 27555 came in at 5:30 P.M. on

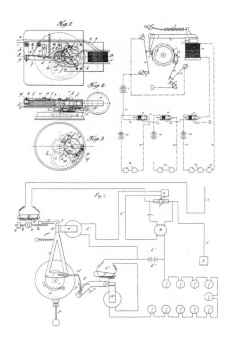

Figure 5.5 Patenting Coordinated Time. Patents on the electromagnetic coordination of time surged around 1905. Here are but a few: Swiss Patent 33700 (upper left) shows a mechanism for the electric resetting of distant clocks (12 May 1905). Swiss Patent 29832 (from 1903) in the upper right illustrates a proposal for the electrical transmission of time. The distant clocks that were to be controlled are visible at the bottom of the wiring scheme. Swiss Patent 37912 (1906), depicted at the bottom of the figure, is one of the earliest approved applications devoted entirely to the radio transmission of time. Such schemes date almost to the first days of radio and were widely discussed in 1905. SOURCE: 33700 (JAMES BESANÇON AND JACOB STEIGER); 29832 (COLONEL DAVID PERRET); AND 37912 (MAX REITHOFFER AND FRANZ MORAWETZ).

7 November 1902 (issued in 1903), offering "An Electric Installation for the Transmission of Time"; Perret floated a similar proposal
in 1904. Mister L. Agostinelli, applying from Terni (patent 29073,
issued 1904) proposed an "Installation with Central Clock for Indicating the Time Simultaneously in Several Places Separated from
One Another, and with Bells for Calling Automatically at Prede-
termined Times." There were patent applications by huge electrical combines like Siemens ("Motherclock Relay," number 29980,
issued 1904) and patents by smaller but still important Swiss firms
like Magneta (patent 29325, submitted 11 November 1903, issued
1904), which manufactured the remotely set clocks gracing the Federal Parliament Building in Bern. A Bulgarian took out a patent in
early 1904 for a mother clock and its electrical secondaries. Applications stacked up in Bern by the dozens.[52] From New York, Stock-
holm, Sweden, London, and Paris, inventors launched their timing
dreams toward the patent office, but it was the Swiss clockmaking
industry that dominated the trade.

During the time that Einstein served as a patent inspector, interest in electrically controlled clock systems heightened. From 1890
to 1900 there were three or four electric time applications each year
(with the exception of two in 1890 and six in 1891). As electric time
transmission grew alongside the telegraph system, coordinated
clocks began playing an ever-increasing role in both public and private sites. The numbers: 1901, eight patents; 1902, ten; 1903, six,
and then in 1904 (*the* peak year from 1889 to 1910) fourteen
patents on electric clocks overcame the hurdles of the patent office.
Numerous others, lost to history, no doubt withered under Einstein's and his colleagues' critical gaze.[53]

All these Swiss chronometric inventions—along with a great
many others related to them—had to pass through the patent office
in Bern and no doubt many of them crossed Einstein's desk.[54] When
Einstein began work there as a technical expert, third class, he was
chiefly charged with the evaluation of electromagnetic and electro-

mechanical patents.[55] At his wooden podium, alongside twelve or so other technical experts, Einstein dissected each submission to extract its underlying principles.[56]

Einstein's expertise on electromechanical devices came in part from the family business. His father Hermann and uncle Jakob Einstein had built their enterprise out of Jakob's patents on sensitive electrical clocklike devices for measuring electrical usage. One of Einstein and Co.'s electrical meters featured prominently in the 1891 Frankfurt Electrotechnical Exhibit, a few pages away from a mechanism (typical of the period) for mounting a backup mother clock to ensure the continued operation of a system of electrical clocks. So close were the electrical measuring systems and electric clockwork technologies that at least one of the Jakob Einstein-Sebastian Kornprobst patents explicitly signaled its applicability to clockwork mechanisms. Conversely, numerous patent applicants proffered devices that applied as much to electricity-measuring systems as to electric clocks.[57]

During Einstein's patent years (June 1902 until October 1909) machines of every sort surrounded him. Sadly, only a few of Einstein's expert opinions survive, rescued from automatic bureaucratic destruction only because the patent process had driven negotiations into court. In one, from 1907, Einstein took aim at a proposal for a dynamo put forward by one of the most powerful electrical companies in the world, the Allgemeine Elektrizitäts Gesellschaft (AEG): "1. The patent claim is incorrectly, imprecisely, and unclearly prepared. 2. We can go into the specific deficiencies of the description only after the subject matter of the patent has been made clear by a properly prepared claim."[58] Description, depiction, claim: these were the building blocks of any patent; to demand their rigorous execution formed the curriculum of Einstein's schooling under Haller.

A second extant opinion by Einstein concerned a patent infringement suit between a German firm, Anschütz-Kaempfe, and an

Figure 5.6 Jakob Einstein & Co. Einstein's uncle and father ran an electrotechnical company that produced, inter alia, precision electrical measuring equipment that shared much technology with that of electric clocks. SOURCE: OFFIZIELLE ZEITUNG DER INTERNATIONALEN ELEKTROTECHNISCHEN AUSSTELLUNG, FRANKFURT AM MAIN (1891), P. 949.

American company, Sperry. Competition to build workable gyroscopic compasses in the early 1910s was ferocious. Metal boats, newly electrified, were terrible sites for magnetic compasses. But in the race to outfit the world's ships and airplanes, Anschütz-Kaempfe suspected the Americans of having stolen their invention. Hermann Anschütz-Kaempfe (the firm's founder) called Einstein, who made short shrift of the Americans' claim that an 1885 patent actually lay behind all their "new inventions." Noting that the 1885 patent described a gyroscope that could not move freely in all three dimensions, Einstein skewered the Americans' claim by pointing out that the older machine could not possibly work accurately in the pitch and roll of a craft at sea. Anschütz-Kaempfe won. Einstein became such an expert on gyrocompasses that he was able to contribute decisively in 1926 to one of Anschütz-Kaempfe's major patents, for which he received royalties until the distributing firm was liquidated in 1938. Significantly, in 1915 the gyrocompass served explicitly as a model for Einstein's theory of the magnetic atom. This cross-talk between machines and theory so riveted his attention that Einstein temporarily put aside his work on general relativity to do a series of collaborative experiments—exceedingly delicate ones—showing that the iron atom indeed functioned like a submicroscopic gyroscope.[59] Patent technology and theoretical understanding were closer than they might have appeared.

A few years later, Einstein made it clear that he treated the very processes of writing and reading as if he were attacking a patent, even when he was treating matters far removed from dynamos or gyrocompasses. In July 1917 his long-time friend, the anatomist and physiologist Heinrich Zangger, wrote Einstein to solicit his judgment of a text Zangger was assembling on medicine, law, and causality. Einstein replied that he liked the "concrete cases," "[b]ut I did not like some of the abstract parts; they often seem to me to be unnecessarily opaque (general) and in the process are not worded clearly and pointedly enough (not every word is placed clearly and consciously). Nevertheless, I understood everything; it may well be possible that my perpetual ride on my own hobbyhorse and the conventions at the Patent Office have driven my standards to exaggerated heights in this regard."[60]

Einstein's fascination with machines and the conventions of the patent office spilled from his workday into the rest of his life. In incessant corresponding about machines with his friends, Conrad Habicht (of the Olympia Academy) and Conrad's brother Paul (an apprentice machine technician), Einstein bandied about ideas for relays, vacuum pumps, electrometers, voltmeters, alternating current recorders, and circuit breakers. Paul, in particular, cooked up one scheme after another, sometimes writing Einstein every two days. On one occasion, having dropped Einstein a detailed letter about a proposed flying machine (a helicopterlike contraption), Paul directly requested advice: "Should I take out a patent immediately? Or publish without a patent or start negotiations without a patent?"[61]

Einstein, too, sought patents. One of his many ideas was for a sensitive electrometer that would measure exceedingly small voltage differences. The little machine (*Maschinchen*) fascinated him in all its aspects, from matters of theory to the nitty-gritty of fabrication.[62] He wrote to one of his collaborators about the gasoline they would need to clean ebonite (vulcanized rubber) parts, and the arrangement required to ensure that the wires would dip properly into mercury beads on the ebonite disk. "I already know from my experiments," he added, "that the thing can be done with the mercury contacts. It is only sometimes time-consuming to put the apparatus in working order, and the mercury spurts out as soon as one turns a bit too fast."[63]

Concerns about instruments, along with experiments on real and imagined machines, held Einstein's rapt attention. Writing to his friends he would switch, without missing a beat, from the rarified-theoretical to the practical-technological. In one letter of 1907, in the midst of telling Conrad Habicht about an article that he was writing on the relativity principle and some current work on the perihelion of mercury, Einstein in the very next sentence was back to the *Maschinchen*. Just about a year later, he told his friend and collaborator Jakob Laub that "In order to test [the *Maschinchen*] for volt[ages] under 1/10 volt, I built an electrometer and a voltage battery. You wouldn't be able to suppress a smile if you saw the magnificent thing that I attached together myself."[64] These efforts on the *Maschinchen* and his later work on the gyrocompass and Einstein–de Haas effect, are but a few examples of Einstein's particular interest in sensitive electromechanical devices that would bridge worlds of electricity and mechanics. Electromagnetic clock coordination proposals were right up his alley — they offered ways to transform small electrical currents into high-precision rotary movements.

Time coordination patents continued to pour into the office. On 25 April 1905 at 6:15 P.M., for example, the office recorded the arrival of a patent application for an electromagnetically controlled pendulum that would take a signal and bring a distant pendulum clock into accord.[65] All such inventions required documentation, including a model, specific drawings, and properly prepared descriptions and claims. Evaluating them was painstaking and often lasted for months.

Sometime in the middle of May 1905 (and we note that Einstein moved outside Bern's zone of unified time on May 15th), he and his closest friend, Michele Besso, had cornered the electromagnetism problem from every angle. "Then," Einstein recalled, "suddenly I understood where the key to this problem lay." He skipped his greetings the next day when he met Besso: "'Thank you. I've completely solved the problem.' An analysis of the concept of time was my solution. Time cannot be absolutely defined, and there is an inseparable relation between time and signal velocity."[66] Pointing up at a Bern clock tower — one of the famous Bern synchronized clocks — and then to the one and only clock tower in nearby Muri (the traditional aristocratic annex of Bern not yet linked to Bern's Normaluhr), Einstein laid out for his friend his synchronization of clocks.[67]

Within a few days Einstein sent off a letter to Conrad Habicht, imploring him to send a copy of his dissertation and promising four new papers in return. "The fourth paper is only a rough draft at this point, and is an electrodynamics of moving bodies which employs a modification of the theory of space and time; the purely kinematic part of this paper [beginning with the new definition of time synchronization] will surely interest you."[68] Ten years of thought about

Figure 5.7 Bern-Muri Map. Michele Besso recalled that when Einstein excit-
edly told him of his realization that time had to be defined by signal exchange,
he pointed to one clock tower in old Bern and to another (the only one) in the
nearby town of Muri. Since it is the only vantage point from which both might
have been visible, Besso and Einstein must have been standing on the hill shown
to the northeast of downtown Bern. SOURCE: MODIFIED FROM SKORPION-VERLAG.

the physics of moving bodies, light, ether, and philosophy had cul-
minated in this little paper.

Obviously time synchronization by the exchange of electromagnetic signals was not all of special relativity, but it was the crowning
step in Einstein's development of the theory. Coordinating clocks
did not, as it had for Poincaré, enter Einstein's reasoning as the physical interpretation of Lorentz's fictional "local time." Far from it.
Einstein began his argument by defining the position of a point relative to rigid measuring rods (a commonplace), but now supplemented with his new *definition* of simultaneity: "If we want to
describe the *motion* of a material point, we give the values of its
coordinates as a function of time. However, we should keep in mind
that for such a mathematical description to have physical meaning,

Figure 5.8 Muri Clocktower (circa 1900). When Einstein gestured toward Muri's only clock tower as he explained to Besso his new time-coordination scheme, this is the structure to which he pointed. SOURCE: GEMEINDESCHREIBEREI MURI BEI BERN.

we first have to clarify what is to be understood by 'time'." With no frame of ether at rest on the basis of which to pick out the one true time, the clock systems of every inertial reference frame were equivalent in the sense that the time of one frame was just as "true" as any other.

In this context, Einstein's paper, completed by the end of June 1905, can be read in a very different fashion from our current standard interpretation. Instead of a wholly abstracted "Einstein philosopher-scientist"—lost in theory while absentmindedly merely earning his keep in the patent office—we can recognize him *also* as "Einstein patent officer-scientist," refracting the underlying metaphysics of his relativity theory through some of the most symbolized mechanisms of modernity. The train arrives in the station at 7:00 P.M. as before, but now, after our long voyage through time and space, we can see it was not just Einstein who was worried about what this means in terms of distant simultaneity. No, determining train arrival times by using electromagnetically coordinated clocks was *precisely* the practical, technological issue that had been racking North America and Europe for the last thirty years. Patents were racing through the system, improving the electrical pendula, altering the receivers, introducing new relays, and expanding system capacity. Time coordination in the Central Europe of 1902–05 was not merely an arcane thought experiment; rather, it critically concerned the clock industry, the military, and the railroads *as well as* a symbol of the interconnected, sped-up world of modernity. Here was thinking through machines.

Einstein brought into his world of principled physics the powerful and highly visible new technology embodied all around him: the conventionalized simultaneity that synchronized train lines and set time zones. A trace of that existing time-coordination system is there to see in the 1905 article itself. Reconsider the coordination scheme with which Einstein begins the paper: An observer is equipped with a clock at the center of the coordinate system. That master clock bolted to space position (0,0,0) determines simultaneity when electromagnetic signals from distant points arrive there at the same local time. But now this standard, centered system no longer appears as an abstract straw man. This branching, radial clock-coordination structure—visible in wires, generators, and clocks, displayed in patent after patent and book after book on timekeeping—*was precisely that of the European system of the mother clock along with its secondary and tertiary dependents.* (See figures 5.9–5.11.) Einstein had brought to that established system the critical gaze of a physicist-patent officer: keep the idea that time had to be defined through a realizable signal-exchange, invoke the absolute speed of light, and cancel the system's dependency on any specific, privileged spatial origin or rest frame for the ether.

With a certain caution, we might attempt to track Einstein's precise train of thought even further into this mid-May 1905 crossing point of the technology and physics of simultaneity. One possibility is this: sensitized to the importance of procedurally based concepts,

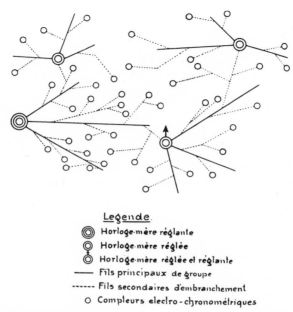

Legende.

◎ Horloge-mère réglante

Ⓧ Horloge-mère réglée

◉ Horloge-mère réglée et réglante

——— Fils principaux de groupe

- - - - - - Fils secondaires d'embranchement

o Compleurs electro - chronométriques

Figure 5.9 Favarger's Time Network. Taking over from Hipp, Favarger led the Swiss company (after 1889, A. Peyer, A. Favarger & Cie.) to a leading position in the world of electrocoordinated clocks, not only in production but also in invention and patenting. Here Favarger depicted a prototypical network of secondary timepieces linked to a master clock. SOURCE: FAVARGER, "ELECTRICITÉ ET SES APPLICATIONS" (1884–85), P. 320; REPRINTED IN FAVARGER, L'ÉLECTRICITÉ (1924), P. 394.

time coordination and its precise, patent applications, and his Olympia Academy critical-philosophical discussions, Einstein could have been fully primed to take up and transform any mention of clock coordination in the context of electrodynamics of moving bodies. Perhaps at some stage Einstein had actually read Poincaré's 1900 article in which the French savant gave his first physical interpretation of Lorentz's (approximate) "local time" as clock coordination by signal exchange in the ether. It is certain that

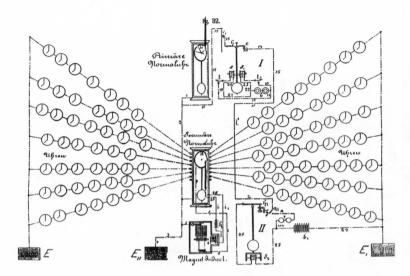

Figure 5.10 Time Telegraph Network. *Another paradigmatic representation of distributed electrical time.* SOURCE: LADISLAUS FIEDLER, *DIE ZEITTELEGRAPHEN UND DIE ELEKTRISCHEN UHREN VOM PRAKTISCHEN STANDPUNKTE* (VIENNA, 1890), PP. 88–89.

sometime before 17 May 1906, Einstein did read Poincaré's 1900 paper — on that day Einstein submitted a paper of his own that explicitly used the contents of Poincaré's article (though *not* local time).[69] Could it be that Einstein had studied or at least seen Poincaré's paper between December 1900 and May 1905? More specifically: had Einstein read and dismissed Poincaré's 1900 ether-based reasoning, while, at some level of awareness, he retained the senior French scientist's provision of a clock-synchronization account of Lorentz's local time? Einstein did not read French easily. But he need not have read Poincaré directly; he could have encountered related ideas (in German) in Emil Cohn's November 1904 article "Toward the Electrodynamics of Moving Systems."[70]

Student of a Strasbourg experimentalist renowned for his work on the speed of sound, Cohn was a successful theorist who had

UNiFiCATiON ÉLECTRiQUE DE L'HEURE DANS UNE GRANDE ViLLE

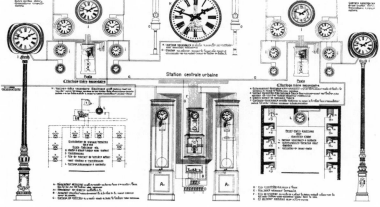

Figure 5.11 Electric Unification. *Favarger wanted to wire the interior of buildings, but more ambitiously the entirety of major urban centers, as this schematic made clear.* SOURCE: FAVARGER, *L'ÉLECTRICITÉ* (1924), PP. 427–28, PLATE 4.

begun his career in the laboratory, measuring magnetism. Even as he turned decisively from bench to blackboard, Cohn relentlessly insisted on the measurable consequences of his work. By the turn of the century Cohn stood as a theorist of repute, joining fellow luminaries to deliver a paper at the December 1900 Lorentzfest. There it seems likely that he would have heard Poincaré's lecture; at the very least he would have had good reason to see the printed version of Poincaré's talk in the published proceedings that contained his own paper. But for our purposes, the intriguing piece of the story is this: in 1904, Cohn, like Poincaré, explicitly introduced clock coordination into his physical definition of local time, and he did so with a former experimentalist's suspicion of purely hypothetical quantities. In this respect he was rather more like Einstein than Poincaré. Unlike Poincaré, Cohn rejected the ether, preferring "the vacuum," and taking local time as given by light-signal coordinated clocks valid for optics if not mechanics.

Again, the unanswerably detailed question: When did Einstein see Cohn's procedural local time? And again, there is little certainty. Only this: sometime before 25 September 1907, Einstein had Cohn's article in hand (that day he wrote the editor of a journal reporting just this, misspelling Cohn's name "Kohn"). On 4 December 1907, the publisher recorded the arrival of Einstein's review paper on relativity that contained this somewhat cryptic footnote-compliment: "The pertinent studies of E. Cohn also enter into consideration, but I did not make use of them here."[71] Again, Einstein would have had to dismiss the vast bulk of Cohn's particular approach to electrodynamics while extracting the idea of clock coordination as relevant to the definition of simultaneity. (Yet another physicist, Max Abraham, had also begun exploring signal-exchanged simultaneity in his 1905 text on electrodynamics, though not early enough for Einstein to have seen it before submitting his article.[72]) The limits of reconstruction are evident, and would be even if it were written in stone that Einstein had seen one of these papers. But the larger goal must stay in view: to understand all we can of the philosophical, technical, and physics conditions that put Einstein in a position to seize clock coordination as the principled starting point of relativity. Of course, we can even sketch possibilities for the seed around which the idea condensed, while avoiding an overwrought attempt to specify exactly what the precipitating seed might have been: a half-remembered line from Poincaré or Cohn encountered in the library, a particular patent application at work, a synchronized Bernese street clock, or a philosophical text batted around the Olympia Academy. Our position is not so different from that of the meteorologists who can provide an excellent account of how a thunderstorm formed by studying the powerful updraft of an unstable column of moist air: but it is not given to know around which bits of dust the initial raindrops precipitated.

Who saw what when? What was deleted, absorbed, provoked by each comment and paragraph? Doling out credit and priority, playing prize committee to the long-deceased is to employ an uncertain history against a futile goal. More important, more interesting is this: in the years before May 1905, simultaneity talk was growing denser among physicists as they grappled with the electrodynamics of moving bodies. Simultaneity procedures lay thicker in philosophical texts, in the cityscape of Bern, along train tracks throughout Switzerland and beyond, in undersea telegraph cables, and in the application in-pile at the Bern patent office. In the midst of this extraordinary material and literary *intensification* of wired simultaneity, physicists, engineers, philosophers, and patent officers debated how to make simultaneity visible. Einstein did not conjure these various flows of simultaneity out of nothing; he snapped a junction into the circuit enabling the currents to cross. Simultaneity had long been in play at many different scales, but Einstein showed how the *same* flashed signal of simultaneity illuminated all of them, from the microphysical across regional trains and telegraphs to overarching philosophical claims about time and the Universe.

Back in 1900 Poincaré had launched his interpretation of time as signal exchange in the midst of his Leiden speech, attributing it, almost as an aside, to Lorentz. Einstein made signal-simultaneity central at every opportunity, from the moment in May 1905 when he stood with Besso gesturing over the hills at the clocks of Muri and Bern. In a wide-ranging review of the relativity theory at the end of 1907, Einstein reiterated the pivotal role that time must play. By his lights, Lorentz's old theory of 1895 had shown, at least approximately, that electrodynamic phenomena would not reveal the motion of the earth with respect to an ether. Michelson and Morley's experiment made clear that even Lorentz's approximate equivalence was not good enough — motion through the ether could not be detected at even higher levels of accuracy. "Surprisingly," Einstein added, "it turned out that a sufficiently sharpened conception of time was all that was needed to overcome the difficulty." Lorentz's 1904 (improved) "local time" was enough to address the problem. Or rather, it solved the problem *if*, as Einstein had done, "'local time' was redefined as 'time' in general." Einstein's view was that this "'time' in general" was precisely the time given by the signal-exchange procedure. With that understanding of time, the basic equations of Lorentz followed. So with this dramatic redefinition, Lorentz's theory of 1904 could be set on the right track, with one further, not-so-small exception: "Only the conception of a luminiferous ether as the carrier of the electric and magnetic forces does not fit into the theory described here." More precisely, Einstein dismissed the idea that electric and magnetic fields were "states of some substance," as the ether advocates would have it. For Einstein electric and magnetic fields were "independently existing things," as freestanding as a block of lead. Electromagnetic fields, like ordinary ponderable matter, could carry inertia. Fields did not depend on the state of an undetectable ether for which Einstein had not the slightest use.

There were many choices open to a physicist wanting to understand this post-1905 controversy about the electrodynamics of moving bodies. Certainly Lorentz and Poincaré loomed large; Einstein's reputation was growing. But there were dozens of ideas vying for attention: the relativity principle, the status of the ether, the absolute speed of light, the changing mass of the electron, the possibility of explaining all mass by electrodynamics. From this swirl, only around 1909 did Einstein's articulation of time rise haltingly, contentiously, but then powerfully. (Even then there were some physicists, such as Ebenezer Cunningham in Cambridge, England, who read Einstein's relativity very differently; he was certainly not the only physicist who while enthusiastic about the new theory did *not* take coordinated clocks to be the main event in the great drama of relativity.)[73]

Clocks First

Göttingen mathematician and mathematical physicist Hermann Minkowski trained his spotlight early and directly on Einstein's clocks. Long before 1905, Minkowski had built his career by applying geometry where no one else had thought to apply it—to the apparently unvisualizable field of the theory of numbers. (It should be said that the younger Einstein resolutely ignored Minkowski, even when he was enrolled in the great mathematician's ETH course.) When Minkowski saw the work of Lorentz, Poincaré, and Einstein, he again looked to geometry. Now Minkowski singled out Einstein's attack on classical time as the key to the puzzle, the key that, in combination with his own formulation of a four-dimensional geometry of "spacetime," opened up a new understanding of physics. Calling the new physics "radical," he even more strongly dubbed it "mightily revolutionary" in his private drafts. In his much studied lecture, "Space and Time," Minkowski made it clear that it was Einstein who had liberated the multitude of physical times from a fictional existence: "time, as a concept unequivocally determined by phenomena, was deposed from its high seat." Minkowski asserted plainly that it was Einstein who had shown that there was no coherent meaning of "time" by itself, only "times," plural and dependent on reference frame.[74]

In pursuing his four-dimensional world, Minkowski drew on Poincaré who, back in 1906, had also considered a four-dimensional spacetime. Poincaré had remarked that amidst all the changes of time and space from one reference frame to another, there was one quantity that remained unchanged. An analogy: Suppose you mark the location of an observatory and nail the map to a board with a single nail driven through the position of your house. Rotate the map 45 degrees clockwise and you change the horizontal distance from your house to the observatory while you simultaneously alter the vertical distance between house and observatory. But obviously rotating the map in this way does nothing to change the actual distance between home and observatory. That is, if the horizontal separation is A, the vertical separation, B, and the distance C, then rotating the map will change A (if the map is turned so that the house and observatory are aligned vertically, for example, there is no horizontal separation). Similarly a rotation will change B, the vertical separation. But rotating the map cannot change the house-to-observatory *distance*, C. Minkowski showed that the relativistic transformations of space and time could be rigorously considered a distance-preserving rotation in the four-dimensional space made up of ordinary space and time. Just as distance in Euclidean geometry remained the same (despite a rotation), so in relativity there was a new distance that remained untouched by the transformations of space and time separately. [Spacetime distance squared] was always equal to [time difference squared] minus [space difference squared].

In a powerful speech in Cologne on 21 September 1908, Minkowski took up the mathematics of Poincaré, but reordered their interpretation in words that instantly seized the imagination of the many physicists: "Henceforth space by itself and time by itself, are doomed to fade away into mere shadows, and only a kind of union of the two will preserve independence." For Minkowski reality lay not in what we could grasp with our ordinary senses (space by itself, time by itself), but in the distances of this four-dimensional fusion of space and time. Invoking a striking imagery of projections and objects in four-dimensional spacetime, Minkowski would have called to mind among his Gymnasium-educated audience Plato's cave scene in the *Republic*. There Plato described how a prisoner, constrained to view but the shadows of objects dancing on the walls, would find it painful to learn to peer directly at the full three-dimensional objects behind the prisoner's head, much less at the light that lit them. Minkowski insisted that in the old physics of "space" and "time," scientists had been similarly misled by appearance. In speaking of space and time separately, physicists were, like Plato's prisoners, contemplating nothing more than shadows projected onto three dimensions. The full and higher reality of a four-dimensional "absolute world" would reveal itself only through the liberation of thought, and specifically through the insights mathematicians could provide.[75]

At first Einstein resisted Minkowski's formulation, seeing in it needless mathematical complexity. But as he ventured deeper into the theory of gravitation, the notion of spacetime proved ever more essential. Meanwhile, all around him, physicists took up Minkowski's idiom as their road into relativity, a path many found more accessible than Einstein's own.[76] A few rejected Minkowski's view that reality lay in four dimensions and that physics itself should be completely reformulated so it described this higher-dimensional world. Among those dubious about a four-dimensional physics figured, ironically, Poincaré himself, who already in early 1907 had preemptively dismissed the utility of such a project:

> It seems in fact that it would be possible to translate our physics into the language of geometry of four dimensions; to attempt this translation would be to take great pains for little profit. . . . [I]t seems that the translation would always be less simple than the text, and that it would always have the air of a translation, that the language of three dimensions seems the better fitted to our description of the world although this description can be rigorously made in another idiom.[77]

As he had done so often, Poincaré pointed out new lands, identified the passes to them, and then chose to make his own stand on terra cognita. Once Einstein began his pursuit of four-dimensional space-time set in non-Euclidean geometry, he never let go.

And again and again, Einstein came back to the problem of time-keeping. In 1910, for example, he once more insisted that time could not be grasped without clocks. "What is a clock?" he asked. "By a clock we understand any thing characterized by a phenomenon passing periodically through identical phases so that we must assume, by virtue of the principle of sufficient reason, that all that happens in a given period is identical with all that happens in an arbitrary period."[78] If there is no reason to think otherwise we should assume the Universe to be constant. If the clock is a mechanism that drives hands around in circles, then the uniform motion of those hands will mark time; if the clock were nothing but an atom, then time would be marked by its oscillations. By itself Einstein's remark about the significance of "clock" extended the series of philosophical investigations of time by Mach, Pearson, or Poincaré himself in "The Measure of Time." Now, however, a clock need not be a macroscopic object at all: it could even be an atom. At the moment Einstein penned these words, he was writing for a scientific publication, but it is not hard to see how these considerations came to count almost immediately as philosophy. Certainly Einstein's time concept was read that way by the new breed of scientific philosophers that included Moritz Schlick and Rudolf Carnap of the Vienna Circle, or their Berlin ally, Hans Reichenbach.

On 16 January 1911, Einstein appeared before the *Naturforschende Gesellschaft* in Zurich. Once again, the increasingly well-known young scientist outlined his reasoning, from the establishment of the Lorentz theory to the starting assumptions of relativity and the procedure for coordinating clocks. Clearly delighting in his example, Einstein explained how "the thing at its funniest" emerged if one imagined a clock—better still, a living clock, an organism—launched into near light-speed round-trip travel. On return, the being would have scarcely grown older, while those remaining at home would have aged through generations. Though previously skeptical, Einstein even tipped his hat toward Minkowski, whose "highly interesting mathematical elaboration" had revealed a method that made relativity theory's "application substantially easier." It was praise that came too late for Minkowski to appreciate; he had died suddenly in 1909. But Einstein now, like Minkowski, began celebrating the appeal of representing "physical events . . . in a four-dimensional space" and physical relations as "geometrical theorems."[79]

Kleiner, Einstein's old thesis examiner and sometime supporter, then took the floor to laud his former, and formerly difficult, student:

> As far as the principle of relativity is concerned, it is being called revolutionary. This is being done especially with regard to those postulates that are uniquely Einsteinian innovations in our physical picture. This concerns most of all the formulation of the concept of time. Until now we were accustomed to view time as something that always flows, under all circumstances in the same direction as something that exists independently of our thoughts. We have become accustomed to imagine that somewhere in the world there exists a clock that categorizes time. At least one thought it permissible to imagine the thing in such a way. . . . It turns out that the notion of time as something absolute in the old sense cannot be maintained, but that, instead, that which we designate as time depends on the states of motion.[80]

By contrast, for Kleiner the relativity concept itself was hardly "revolutionary"; it was a "clarification," perhaps, but not something "fundamentally new." If the good professor Kleiner mourned anything in Einstein's physics, it was the loss of the ether. True, he conceded, the concept had grown more and more incomprehensible. But without it wouldn't there be "propagation in a medium which is not a medium"? Worse yet, didn't the abandonment of the ether leave us with formulae that lacked any "mental image"? In reply, Einstein allowed that the ether might, in the age of Maxwell, have had "real value for intuitive representation." But the value of the ether concept vanished when physicists gave up trying to picture ether as a mechanical entity with mechanical properties. After the demise of the truly intuitive ether, the notion had become for Einstein nothing but a burdensome fiction.

That January day, just about every speaker directly or indirectly brought up Minkowski's "Space and Time" speech, which clearly had paved the way for a wider acceptance of Einstein's theory. Einstein's exchanges with one member of the audience (a 1904 graduate of the University of Zurich) threw the status of the theory into relief:

> **Dr. [Rudolf] Lämmel:** Is the world picture resulting from the conceptions of the relativity principle an inevitable one, or are the assumptions arbitrary and expedient but not necessary?
> **Prof. Einstein:** The principle of relativity is a principle that narrows the possibilities; it is not a model, just as the second law of thermodynamics is not a model.
> **Dr. Lämmel:** The question is whether the principle is inevitable and necessary or merely expedient.
> **Prof. Einstein:** The principle is logically not necessary: it would be necessary only if it would be made such by experience. But it is made only probable by experience.

For Poincaré, too, principles were made probable by experience, but principles were precisely what was expedient; they could be held against the grain of experience only at the cost of immense inconvenience. "Principles," Poincaré had famously written in *Science and Hypothesis*, "are conventions and definitions in disguise." For Einstein, principles were more than definitions, they were pillars, supports of the structure of knowledge. And this despite the circumstance that our knowledge of principles could never be certain; our hold on them was necessarily provisional, only probable, never forced by logic or experience.

For Ernst Meissner, then a Privatdozent at ETH in physics and mathematics, Einstein's work on time presented a model for a vast and critical reassessment of every concept of physics. Each one would have to be interrogated to identify those that remain invariant when one switches frames of reference:

> **Meissner:** The discussion has shown what is the first thing to be done. All physical concepts will have to be revised.
> **Einstein:** The main thing now is to set up the most exact experiments possible in order to test the foundation. In the meantime, all this brooding is not going to take us far. Only those consequences can be of interest that lead to results that are, in principle, accessible to observation.
> **Meissner:** You have brooded over this, and discovered the magnificent time concept. You found that it is not independent. This must be investigated for other concepts as well. You have shown that mass depends on the energy content, and you have made the concept of mass more precise. You did not carry out any physical investigations in the laboratory—you were brooding instead.[81]

Ah yes, Einstein replied. But think of the fine predicament in which that brooding about time has put us.

Einstein's revision of time seized the attention of some of his most illustrious contemporaries. Max von Laue declared in his 1911 text on the relativity principle that it was the "groundbreaking work" of Einstein that had, with the single stroke of his radical critique of time, solved the puzzle of Lorentz's real but undetectable ether.[82] Max Planck, the dean of German physics (who had introduced the quantum discontinuity into physics), went further. Speaking in 1909 at Columbia University in New York, he told the assembled: "It need scarcely be emphasized that this new view of the concept of time makes the most serious demands upon the capacity of abstraction and the imaginative power of the physicist. It surpasses in boldness everything achieved so far in speculative investigations of nature, and even in philosophical theories of knowledge: non-Euclidean geometry is child's play in comparison."[83] Planck's words lofted Einstein's reputation, broadcasting Einstein's own sense that time was the pivot point of his work. The relativity of simultaneity, Einstein remarked soon afterward, "signifies a fundamental change in our concept of time. [It] is the most important, and also the most controversial theorem of the new theory of relativity."[84]

Emil Cohn, who had been thinking about light-coordinated clocks at least as early as 1904, returned to the question in 1913 in a brief, popular book (*The Physical Aspect of Space and Time*) to confront simultaneity in a thoroughly Einsteinian way (he referred to the "Lorentz-Einstein Relativity Principle," no mention of Poincaré). Including a photograph of a wire and wood model of clock coordination, Cohn drew dozens of clocks and rulers to emphasize at each step of his argument that the Einsteinian kinematics was physical, procedural, and altogether visualizable in terms of coordinated public clocks: "The synchronization of a Strasbourg and a Kehl clock (that have been previously checked to run at similar rates) can and ought to be set in this way: Strasbourg sends at time 0 a light signal to Kehl, which is reflected; it gets back to Strasbourg at time 2. The clock in Kehl then is correctly set if, at the moment that the signal arrives there, its clock shows time 1 (and if not it should be corrected to do so)." Einstein liked Cohn's presentation and said so in print.[85]

Not for a moment did Einstein himself stop "brooding" about time; in 1913 he too published a new and strikingly simple argument for the relativity of time. Imagine, he said, that two parallel mirrors make up a "clock" where each tick is defined by the traversal of a burst of light from one mirror to the other (figure 5.12*a*). Now suppose that this light clock is moving to the right (figure 5.12*b*). To the stationary observer, the up-and-down motion of the flash of light appears to make a sawtooth pattern, much the way the ball of a running basketball player would follow such a trajectory as seen by the spectators. Here's the point: the inclined trajectory of the moving clock (as seen by the stationary observer) is obviously longer than the perpendicular trajectory of the rest observer's own clock. But by assumption the speed of light is the same in *every* frame of reference, so the angling trajectory of light travels at *c*. (This is *not* true for the basketball case, since the spectators would see the incline motion of the ball to be faster than the simple up-and-down motion seen by the player.) Since light has further to travel along the incline than it does along the perpendicular, it takes longer (*D* is greater than *h*). It follows that one tick for the moving observer (which appears to the stationary observer as following an incline) is registered as taking more than one click for the stationary one (which goes straight up and down).[86]

So as far as the stationary frame is concerned, everything that takes place in the moving frame runs slowly. However Einstein presented his theory, the core lesson was the same: Absolute time was finished. In its place he offered a simple, practical procedure: Synchronize clocks by the exchange of light. Everything else in the theory followed from it alongside the fundamental assumptions of relativity and the absolute speed of light.

Radio Eiffel

When center-issued electromagnetic signals arrived at distant points, whether in the next room or a hundred kilometers distant, it was not only Einstein and Poincaré who *defined* them as simul-

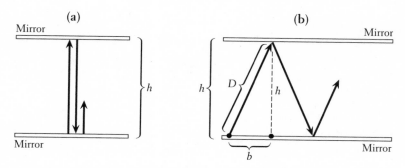

Figure 5.12 Einstein's Light Clock (1913). (a) In this simplest of all explanations of the dilation of time, Einstein imagined two parallel mirrors with a
pulse of light reflecting between them, each traversal constituting a "tick." If
a clock like this flew by an observer "at rest," the observer at rest would see the
pulse following a sawtooth pattern. (b) Each angled traversal would follow a
longer (diagonal) path than in the straight up-and-down path of a similar
clock in the rest frame. Since light travels at the same speed in every frame of
reference, Einstein concluded that a tick in the mirror frame would be measured as taking longer than a tick in the rest frame. Therefore, the rest observer
must conclude, time runs slow in the moving frame of reference.

taneous. Not at all. On the basis of the exchange of electrical signals, railroad planners scheduled trains, generals roused troops,
operators telegraphed business deals, and geodesists drew maps.
Indeed, precisely during Einstein's early patent office years, prepa-
rations were being made to send time coordination signals by radio
waves—the American Navy began experimenting with low-powered
time signals from New Jersey in September 1903, with the order
coming down to broadcast from Cape Cod (Massachusetts) and
Norfolk (Virginia) in August 1904. Nor was radio time just a matter
for the Americans. There was an intense burst of activity surrounding radio coordination systems in 1904 both in Switzerland and in
France as workers tested, developed, and began deploying new
radio time systems. The director of the French journal *La Nature*
himself took up his pen to record new developments in the distrib-

ution of time by wireless. Reporting on experiments conducted at the Paris Observatory, he noted that with the aid of a chronograph, distant synchronization now appeared possible to within two- or three-hundredths of a second. Wireless technologies promised to distribute time everywhere in Paris and its suburbs, supplanting not only the antique steam system but also the cumbersome land lines of telegraphically communicated electric time. Radio time had advanced science through more precise determinations of longitude; now radio would free time from the physical burden of wire. At last, simultaneity could be broadcast to ships at sea and even into "the ordinary household."[87] Patents soon began arriving in Einstein's office with schemes for radio time synchronization.[88]

Wireless time was all the rage in those first years of the twentieth century, and France pushed the new technology hard. Poincaré played a pivotal role through his popular and technical publications on radio, but even more powerfully behind the scenes. As he had so many times before, Poincaré crossed back and forth between abstract considerations of the status of electromagnetism and the practical exigencies of putting radio technology to immediate use. That continuing engagement with the theory and practice of communication no doubt facilitated his 1902 professorial appointment at the Ecole Professionelle Supérieure des Postes et Télégraphes. When, that same year, he contributed an article on wireless telegraphy to the yearbook of the Bureau of Longitude, he began by reviewing Hertz's famous 1888 experiments that first demonstrated the existence of radio waves. But Poincaré immediately plunged himself into practical matters, asking: How could the new "Hertzian light" supplant optical telegraphy by reaching farther and diffracting around the curvature of the earth? How could radio penetrate fog that would block visible light? Could new kinds of antennae direct and concentrate radio waves? Poincaré was as willing to attack the details of nickel-silver coatings as he was to reflect on uses of radio to avoid ship collisions in bad weather.

Figure 5.13 Wireless Time. *In 1904–05 numerous groups were experimenting with the wireless transmission of time. The American Navy was one of the first, but others were pursuing the same goal. In this figure from a widely circulated journal in France, the transmitter, receiver, and operator were all shown.* SOURCE: BIGOURDAN, "DISTRIBUTION" (1904), P. 129.

As we have seen, the French had radio worries of a geopolitical sort after the stinging demonstration of British cable monitoring and cutting capacities during the late 1890s. These were concerns that Poincaré very much shared with other members of the French administrative elite. One French diplomat warned the French Colonial Union that, as long as Britain maintained its telegraph cable monopoly, the French simply could put no faith in the confidentiality of their State messages. Using a lantern slide, he projected for his concerned listeners a dramatically colored worldmap where they could see the desperate situation for themselves. Drawn in blue were the handful of short French cables binding North Africa and France and a single long line to the United States. "Look, now, at the immense development of the red lines: they extend everywhere and encircle the entire world in a veritable spider's web. These red lines mark the network of the English telegraphic companies."[89] Given the instabilities marked by anticolonial uprisings and conflicts threatening the peace between France and Britain, the situation was dire. In this charged atmosphere it is no surprise that for Poincaré security was a key feature of the new wireless: "Optical and Hertzian telegraphy have a common advantage over ordinary telegraphy: in times of war, the enemy cannot interrupt communication." But while the enemy would have to be in the proper position to intercept a light signal, broadcast signals could be captured much more widely, even interfered with by the counterbroadcasting of noise. "We remember that Edison threatened his European competitors, that he would disturb their experiments if they wanted to experiment in America." Back and forth Poincaré oscillated between the pure and the practical, like the spark in a radiotransmitter: primary and secondary circuits were his subjects, but so was the security of French diplomatic communication.[90]

In a search for ever-higher sites on which to post antennae, boosters of the nascent French radio service had already begun to eye the Eiffel Tower, the fate of which remained very much undecided in 1903. Eleuthère Mascart wrote Poincaré from the Bureau of Meteorology with a plea for help in getting the Minister of War to save the tower (and therefore the station) from being dismantled. The great tower was, he argued, a military asset, not only for optical telegraphy but also for the nascent wireless experiments that had already begun. Surely Poincaré would have the ear of the Minister of War—would he try to save the tower on grounds of national defense? Meanwhile, Gustave-Auguste Ferrié, Polytechnician, engineer, and army captain, joined forces with Gustave Eiffel; in 1904 they succeeded in getting Eiffel's tower designated a station of the French radio service. Clinching their success was the army's much-publicized triumph with radio when, in 1907, Ferrié trundled radio equipment into battle on horse-drawn carts so that French forces fighting Moroccan rebels could communicate with their commanders in France.[91]

Given his position in the Bureau of Longitude, in the scientific community, and by then more broadly within the French intellectual elite, Poincaré's radio-time ambitions for the tower carried weight. Largely at his behest, in May 1908 the Bureau urged the establishment of a radio time signal from the Eiffel Tower that could be used to determine longitude anywhere the signal could be received. Backing from the army came easily. In the winter of 1908, the French government launched an interministerial commission to control the new radio technology, appointing Poincaré at its head. The Minister of War concurred, releasing the funds. Wireless simultaneity had become a military, as well as a civilian, priority.[92]

With Poincaré presiding, the seventh meeting of the commission assembled on 8 March 1909. The director of the Paris Observatory was there; so was (the newly promoted) Major Ferrié, as well as engineers from various ministries including the Navy. Explaining the situation to the assembled, Ferrié divided time signals into two classes. A first, rough signal, accurate to about a half-second, could be used by navigators at sea. Such navigation pulses could be initiated by using signals sent by wire from the observatory. Then there were the "special" signals for high-accuracy geodesy. These would need to be crafted more carefully to achieve a precision of one-hundredth of a second.[93]

If you were a radio operator in one of the colonies straining for your precise longitudinal relation to Paris, here's what you would do once you had tuned in Paris. Eiffel Tower station would broadcast a signal once every 1.01 seconds; you would listen for the broadcast pulses beginning before midnight, Paris time. At the same time, you would set your local clock beating a short, crisp tone once every second on the second (local time). By convention you knew that the pips would begin from the Eiffel Tower at midnight Paris time, and so occur at a Paris time of 12:00:00.00, 12:00:01.01, 12:00:02.02, and so on. By counting the number of Eiffel Tower pips before the Eiffel and local beeps coincided, you could synchronize the clocks. For example, if your local whole-second tone first coincided with the Eiffel tones on the tenth Eiffel tone, then you would know that at the Eiffel tower it was midnight plus ten pips (12:00:10.10 seconds). You know the local time of the coincidental beat simply by checking your own clock. So you would subtract Eiffel Tower time (12:00:10.10 seconds) from your local time to get the longitude difference between your radio station and that great symbol of Parisian modernity. By March 1909, Poincaré's commission had a plan for sending precision time signals by wireless.

The next month, Poincaré traveled to Göttingen to deliver a series of lectures on pure and applied mathematics. For the first five of these presentations, Poincaré presented his technical work in German. But at the last meeting he explained to the assembled that this time, without the crutch of equations, he would return to his native tongue. The subject was the "new mechanics." Looking around, he told those gathered that the seemingly imperishable monument of Newtonian mechanics was, if not yet quite flattened, powerfully shaken. "It has been submitted to the attacks of the great destroyers: you have one among you, M. Max Abraham, another is the Dutch physicist M. Lorentz. I want to speak to you . . . of the ruins of that ancient edifice and the new building that one wants to raise in their place." Poincaré then asked, "What role does the principle of relativity play in the new mechanics?" He continued, "We are first led to talk about apparent time, a very ingenious invention of the physicist Lorentz." Picture, he urged his audience, "meticulous observers such as hardly exist. They demand a clock setting of an extraordinary exactitude; to be [accurate to] not one second, but a billionth of a second. How could they do it? From Paris to Berlin, A sends a telegraphic signal, with a wireless if you want, to be altogether modern. B notes the moment of reception, and that would be, for the two chronometers, the [zero] point of time. But the signal takes a certain time to go from Paris to Berlin, it can only go with the speed of light; B's watch will therefore be slow; B is too intelligent to not have realized this; he will take care of this drawback." Observers A and B solve the problem the way any two Bureau of Longitude telegraphers would—by crossing signals. A sends a time signal to B, and B to A.[94] It is just the way the bureau has been doing business for decades—sending signals back and forth by cable between Paris and Brazil, Senegal, Algeria, America. Or, for that matter, as Poincaré would soon help make possible, by way of wireless between the "altogether modern" Eiffel Tower and Berlin.

At 2:30 P.M. on Saturday, 26 June 1909, Poincaré and his commission gathered at the Eiffel Tower to inspect the experimental station. Ship's Captain Colin described and explained the apparatus, summing up the latest work in radio-telegraphic inventions and pointing out and distributing a report from the American Navy on its recent radio synchronization of their ships' clocks. Then came his summary of the ever-accelerating range of the tower itself: over the last days Colin's troops had successfully received signals in Villejuif, 8 kilometers from the tower; in Mehun, 48 kilometers away; on up through a mid-June triumph when engineers captured a transmission 166 kilometers from the Champ de Mars. Even greater range seemed possible. Shipboard experiments had been successfully in operation since 9 June. "The commission, immediately on ending the explanations of Monsieur Colin, set the apparatus in operation." Ammeter, wavemeter, and receiver at the ready, Poincaré's commission elatedly witnessed a flawless (pure and stable) broadcast of time. Poincaré pressed the Chamber of Deputies for commercial radiotelephone services and for immediate funding to make the Eiffel Tower into the greatest time synchronizer in the world. Approval came on 17 July 1909.[95]

Exactly one week later, on 24 July, Poincaré put the finishing touches on his opening speech for the French Association for the Advancement of Science in Lille. In early August, when he entered the town's Grand Théâtre to deliver that address (modified from the one he had given in Göttingen) and receive the Grande Médaille d'Or, he found the city's elite gathered to hear him speak on the new physics. Again he underlined the importance of the relativity principle, the centrality of Lorentz's "ingenious" local time, and the necessity of telegraphic time coordination making use of the "altogether modern" wireless. Poincaré then introduced, as he had before, "another hypothesis" (that is, one beyond the hypothesis of the principle of relativity and the hypothesis of "apparent time"). This third supposition was Lorentz's contraction: an idea "more surprising, much more difficult to accept, which disturbs greatly our current habits." An object in motion through the ether underwent a contraction along its line of motion. As a result of its orbit around the sun, the earth's sphere would be compressed in the direction of its motion by about 1/200,000,000 of its diameter. Nonetheless, Poincaré emphasized, that in the moving observer's frame of reference, the slowing of "local time" and the shortening of "apparent length" so precisely compensated for each other that there would be no way for the moving observer to discover that he was, in fact, moving.[96]

Poincaré's is a description of the world that resembles Einstein's in what the moving observer sees, yet differs in how that circumstance is explained. Here, in August 1909, Poincaré maintained his (ever-less physical) ether, whereas Einstein polemicized at every turn against what he considered an antiquated and redundant entity. Poincaré introduced the Lorentz contraction as a separate hypothesis, Einstein derived it from his definition of time. Poincaré guarded Lorentz's venerable "local time" and "apparent lengths" though his and Lorentz's uses of the terms were not identical: Poincaré treated apparent lengths and times as observable, Lorentz continued to treat them as fictional. For Einstein there was simply "a time for a particular frame of reference," the time of one frame was as "true" or "real" as the time of another. No fictions. No ether. Nothing to be "explained" about the inability of an observer to detect his or her constant motion. No daylight between "true" and "apparent." As for what could be seen: for years Poincaré had been as clear as Einstein—for *any* constantly moving observer *all* the phenomena are "well in accord with the principle of relativity."[97]

From 1908 to 1910, Poincaré's engagement with electromagnetic simultaneity crossed and recrossed between relativity and the still-new radio technology. After recovering from a surge of Seine flood waters into their radio headquarters, the French Army's time team at the Eiffel Tower began broadcasting simultaneity on 23 May 1910, its distinctive tones available to be plucked out of the ether (so to speak) from Canada to Senegal.[98] At first the signals followed Paris time; only the next year, on 9 March 1911, did France agree to reset its (and Algeria's) clocks by the 9 minutes and 21 seconds that riveted them to Greenwich. Charles Lallemand, who had fought alongside Poincaré in the various time and longitude campaigns (Quito, decimal time, time zones), saw the opportunity to put time unification into practice. Clocks, radios, and map making at last converged.

Even before the Eiffel Tower station opened for business, the French longitude men had begun radio-correcting their maps, starting with Montsouris, Brest, and Bizerte, drawing up plans for use by military telegraphy and planning radio time-coordination to map French colonies. Soon they formed a collaboration with their American counterparts to exchange signals between the Eiffel Tower and the Arlington, Virginia, transmitter with sufficient precision to correct for the signal's transoceanic time-of-flight. It was a full-dress realization of the light-signal synchronization that Poincaré had written first into his metaphysics of simultaneity and then into his physics of local time. Popular journals took notice: "Although radio signals travel through space at approximately the velocity of light, there is a slight but appreciable loss of time . . . in starting such a signal on its journey . . . and in receiving it at the other end." When the Americans formulated experimental protocols in 1912, they discovered that the French had already solved the problem, using the coincidence method Poincaré's commission had advanced several years earlier. "Here," one American reporter wrote, "was a beautiful solution."[99] Wireless had made world-synchronization possible: in all directions, over vast distances, with an essentially limitless precision. Lallemand and his longitude allies hoped to coordinate Paris with other transmitters supplied with time signals by an aristocratic consortium of observatories that would "avoid all appearances of a national time." Ministers, observatory directors, and longitude authorities agreed: the French ideal would achieve a rational, coordinated, international system with France clearly, if unautocratically, at the helm.[100] Once again, the triumph would be of a convention in all its senses, building on the long series of international conferences and agreements establishing the meter, ohm, prime meridian, and the abortive attempt to create rational universality from the decimalized second.

While French scientists wired the Eiffel Tower to align clock hands across Europe, the British kept silent, refusing to build a time transmitter of their own. According to one historian of Greenwich, the Imperial telegraphers and astronomers saw the French and other foreign radio-time services as useful in times of peace (the British quickly installed receivers at Greenwich). In war, they reckoned, no one would broadcast time.[101] Such a calculated, pragmatic stance was certainly consistent with the way the British built and controlled their international cable network. For his part, Poincaré had hardly been alone in his patriotic concerns about radio secrecy. He had emphasized the point in his very first article on radio technology in 1902, and throughout the interministerial meetings he oversaw, the delicacy of "secret correspondence" was a subject returned to often as the commission prepared for the 1912 international conferences on time and radiotelegraphy. Repeatedly the French ministerial representatives warned that German and British radio transmitters had outflanked and outclassed the French in colonial Africa.[102] Powering up the Eiffel Tower as the beacon of world time (and zero point of communication with the French Empire) was therefore an effort at once practical and symbolic, military and civilian, nationalist and internationalist. Fourteenth-century villagers had mounted clocks on bell towers to reign over all who could hear them; Poincaré rang the Eiffel Tower radio-clock through the ether to echo French scientific authority across the world.

But whether by telegraph line or by wireless, centered systems were the temporal-physical glory of Europe's Great Powers. It was the unified German empire that von Moltke wanted, made corporal through the grand *Primäre Normaluhr* at the Schlesischer Bahnhof in Berlin or the baroque and elegant *horloge-mère* of Neuchâtel. It was Eiffel's engineered modernity turned high-tech radio timepost; it was Britain's cable network sliding its copper tentacles from Greenwich around its cross-continental colonies. It was the American Navy's powerful radio transmitter directing ships on the high seas and fixing the positions of the Americas' land stations, at the height of big-stick diplomacy.

The widening gyre of technological, symbolic, and abstract physics spiraled outward. Telegraphers, geodesists, and astronomers *understood* the Poincaré-Einstein clock coordination by means of quite literal, everyday wired (and wireless) clock coordination. At the Ecole Supérieure des Postes et Télégraphes, where Poincaré

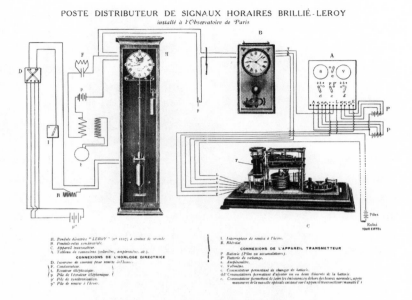

POSTE DISTRIBUTEUR DE SIGNAUX HORAIRES BRILLIÉ-LEROY
installé à l'Observatoire de Paris

Figure 5.14 Eiffel Station Schematic. *Military radio saved the Eiffel Tower, although Poincaré's efforts to use it as a vast antenna in a time-transmission system eventually produced a facility available to both civilians and the armed forces. This figure depicts the master clock (in the Paris Observatory) and its links to the Tower's radio transmitter.* SOURCE: L. LEROY, "L'HEURE" (N.D.), PP. 14–15.

had taught since 1902, the telegraphic and wireless networks were never *only* metaphorical. They were the company business. On 19 November 1921, physicist Léon Bloch explained the meaning of time in a major lecture on relativity theory, using a technology that his audience of students and faculty knew like the backs of their hands:

> What do we call time on the surface of the earth? Take a clock that gives astronomical time—the mother pendulum of the Observatory of Paris—and transmit that time by wireless to distant sites. In

Figure 5.15 Eiffel Radio Time (circa 1908). The Eiffel Tower radio station was located in these unprepossessing shacks at the foot of the structure. SOURCE: BOULANGER AND FERRIÉ, *LA TÉLÉGRAPHIE SANS FIL ET LES ONDES ÉLECTRIQUES* (1909), P. 429.

what does this transmission consist? It consists of noting at the two stations that need synchronization, the passage of a common luminous or hertzian signal.[103]

By the time of Bloch's lecture, clock coordination by exchange of electromagnetic waves was a *practical* routine. For a full decade Post and Telegraphs, the Bureau of Longitude, and the French Army had been routinely correcting for signal time in their myriad of long-distance synchronizations.[104] Einstein and Poincaré's time coordination was born in a world of machines and it clearly was received that way, not just in France. In Germany it was the experimenter-turned-theorist Cohn who had seized light-signal synchronization shortly after Poincaré and lost little time after Einstein's 1905 paper in using pictures of clocks and models to publicize the new simultaneity. In Cambridge, England, it was the experimenters (not the more mathematical theorists) who first seized on the procedures of clock coordination. For American theoretical physicist John Wheeler, the identification of theory with mechanisms and devices tracked across the whole of his career, from early radio and explosives through his apprenticeship to engineers and engineering physicists during World War II. When he and Edwin Taylor wrote their widely used text *Spacetime Physics* in 1963, they too invoked a universal machine and displayed it at the beginning of their book (see figure 5.16).

Machines tied clocks and maps ever closer together. At the beginning of World War II, MIT scientists used improvements in timing to develop the Long Range Aid to Navigation (LORAN) system that guided Allied ships across the Pacific. Postwar projects by the American Navy and Air Force tumbled out one after the other under such names as "Transit" and "Project 621B." As the Cold War intensified, the American military demanded ever more precise locating systems to aim intercontinental ballistic missiles from shifting platforms and to guide soldiers through the unmarked jungles of southeast Asia.

During the 1960s, American defense planners turned satellites into radio stations that would beam timed signals to earth. More accurate and stable timepieces drove these orbiting transmitters, pinging time at first from quartz crystals and later from the cesium oscillations of space-based atomic clocks. By the time the $10 billion Global Positioning Satellite (GPS) system was up and functioning in the 1990s, its twenty-four satellite-based clocks ticked with a precision about which Poincaré only fantasized in his Göttingen or Lille lectures of 1909: 50 billionths of a second per day providing a resolution on the earth's surface of fifty feet. In a certain

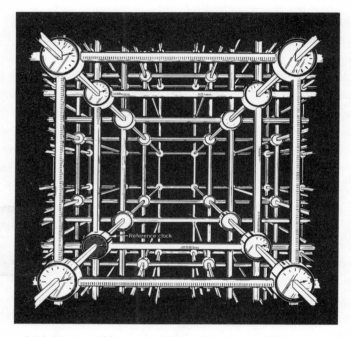

Figure 5.16 Lattice of Space and Time. *Lost in many discussions of relativity are the machinelike procedures by which time and space coordinates were to be mapped. These are eminently visible in this fanciful machine depicted in the relativity textbook by Edwin Taylor and John Wheeler.* SOURCE: TAYLOR AND WHEELER, *SPACETIME PHYSICS* (1966), P. 18.

sense, the system resembled Eiffel Tower time: GPS also used a kind of coincidence method to synchronize clocks. But now the satellites broadcast a string of pseudo-random numbers (that is to say random enough for these purposes)—six thousand billion digits. The receiver then matched this string against its own, identical, internally stored set. By determining the offset between the two series, the logic circuits of the receiver could determine the time difference, and knowing the speed of light, the receiver's distance to the satellite. If the receiver was already synchronized, only three satellites would be needed to fix the receiver's position in three-dimensional space, but since the mobile ground receiver would normally not have the correct time, a fourth satellite (to set the time) was required.

In a trading zone of engineering-philosophy-physics, relativity had become a technology, one that swiftly displaced traditional surveying tools. In fact, by processing the data after the fact and using GPS measurements of a known position to pinpoint tran-

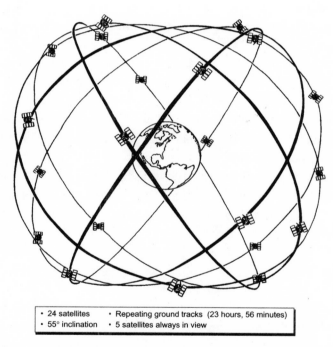

| • 24 satellites | • Repeating ground tracks (23 hours, 56 minutes) |
| • 55° inclination | • 5 satellites always in view |

Figure 5.17 Global Positioning System. Not unlike Poincaré's time-transmitting Eiffel Tower, the late-twentieth-century GPS satellites provided precision timing (and therefore positioning) for both civilian and military users. Built into this orbiting machine were software and hardware adjustments required by Einstein's theories of relativity. The result is a planet-encompassing, $10 billion theory-machine. SOURCE: RAND CORPORATION, RAND MR614-A.2.

sient errors in the system, early twenty-first-century surveyors could use the GPS to determine a second, unknown location within millimeters. So accurate had the system become that even "fixed" parts of the earth's landmass revealed themselves to be in motion, an unending shuffle of continents drifting over the surface of the planet on backs of tectonic plates. In place of "absolute continents," earth scientists demanded a new universal coordinate system, one unattached to any particular surface feature but instead rotating, in the imaginative eye of science, in silent coordination with the interior of the planet. GPS was soon landing airplanes, guiding missiles, tracking elephants, and advising drivers in family cars.

For all these purposes relativistic time coordination was deep in the machine. According to relativity, satellites that were orbiting the earth at 12,500 miles per hour ran their clocks slow (relative to the earth) by 7 millionths of a second per day. Even general relativity (Einstein's theory of gravity) had to be programmed into the system. Eleven thousand miles in space, where the satellites orbited, general relativity predicted that the weaker gravitational field would leave the satellite clocks running fast (relative to the earth's surface) by 45 millionths of a second per day.[105] Together, these two corrections add up to a staggering correction of 38 millionths (that is, 38,000 billionths) of a second per day in a GPS system that had to be accurate to within 50 billionths of a second each day. Before the first cesium atomic clock launch in June 1977, some GPS engineers were sufficiently dubious about these enormous relativistic effects to insist that the satellite's atomic clock broadcast its time "raw." Its relativity-correction mechanism idled onboard. Down came the signal, running fast over the first twenty-four hours almost precisely by the predicted 38,000 billionths of a second. After twenty days of such gains, ground control commanded the frequency synthesizer to activate, correcting the broadcast time signal.[106] Without that relativistic correction, it would have taken less than two minutes for the GPS system to exceed its allowable error. After a single day, satellites would have been raining erroneous positions, skewed by some six miles, onto the earth. Cars, bombs, planes, and ships would have veered wildly off course. Relativity — or rather relativities (special and general) — had joined an apparatus laying an invisible grid over the planet. Theory had become machine.

Imitating historic precedents, symbolic and physical protests against globalized, instrumentalized time were not far behind. In this case one group of protesters explicitly wanted to object concretely to the use of GPS in precision weapons, counterinsurgency warfare, police actions, and nuclear war planning. In the predawn hours of 10 May 1992, two activists from Santa Cruz, California, disguised themselves as Rockwell International workers and broke into the clean room at Seal Beach, California, where the company was readying NAVSTAR GPS satellites for the air force. Slamming an axe into one completed satellite sixty times, they caused nearly $3 million worth of damage. As they approached a second satellite, Rockwell security personnel seized them at gunpoint and turned them over to the police. Dubbing themselves the Harriet Tubman–Sarah Connor Brigade (linking the heroine of the underground railroad to the heroine of "Terminator 2: Judgment Day"), the two militants pleaded guilty and served about two years in jail.[107] In 1996, the FBI reported that the Unabomber, known for his eighteen-year bombing campaign against scientific figures, modeled himself on another temporal anarchist in Joseph Conrad's *Secret Agent*, which he had read a dozen times.[108] Around the axis of global time, fictional, scientific, and high-tech time machines (and their opponents) crossed and recrossed.

Beating overhead in church spires, observatories, and satellites, synchronized clocks have never stood far from the political order — not in the 1890s, not in the 1990s. Poincaré's and Einstein's universal time machines linked technology, philosophy, politics, and physics. Or perhaps we should put it differently: these synchronizing time machines never functioned entirely in the arcane abstract or in the mutely material. The coordination of time is inevitably abstract-concrete.

Turn-of-the-century Europe and North America were criss-crossed with lines of coordination: webs of train tracks, telegraph lines, meteorological networks, and longitude surveys all under the watchful, increasingly universal clock system. In this context, the clock coordination system introduced by Poincaré and Einstein was a world machine: a vast, at first only imagined, network of synchronized clocks that by the turn of the next century had metamorphosed from networks of submarine cables hauled by schooners to a microwave grid broadcast from satellites. There is a sense in which Einstein's special theory of relativity has always been a machine, an imaginative one to be sure, but one suspended in a constantly evolving *real* skein of wires and pulses that synchronized time by the exchange of electromagnetic signals.

Such a *technological* reading of this most *theoretical* development suggests one final observation. It has long struck scholars that the style of Einstein's "On the Electrodynamics of Moving Bodies" does not even look like an ordinary physics paper. There are essentially no footnotes to other authors, very few equations, no mention of new experimental results, and a lot of banter about simple physical processes that seem far removed from the frontiers of science.[109] By contrast, pick up a typical issue of the *Annalen der Physik* and a very different form appears in nearly every article, characterized by the standard point of departure: an experimental problem or a calculational correction. Typical physics articles were and are filled with references to other papers; Einstein's article does not fit this mold. It could be that Einstein's youthful arrogance had simply taken over, perhaps he had dispensed with the niceties of footnotes, altered the form of the usual introduction, and redesigned the typical conclusion as an idiosyncratic matter of individual taste. Certainly self-confidence is what Albert Einstein lacked least.

But read Einstein's contribution through the eyes of the patent world and suddenly the paper looks far less idiosyncratic, at least in style. Patents are *precisely* characterized by their refusal to lodge themselves among other patents by means of footnotes. If you aim to demonstrate the utter originality of your new machine (and upon originality hinges the patent), you can hardly do worse than shower the inspector with a storm of footnotes to prior work. In the fifty or so Swiss electric clock patents granted in the years around 1905, for example (they are typical), there is not a single footnote either to another patent or to a scientific or technical article.[110] This comparison does not *prove*, of course, why Einstein did not cite others in that first paper. But it may help make sense of why a young patent officer in a hurry may not have felt impelled to situate his work in the matrix of papers by a Lorentz, a Poincaré, an Abraham, or a Cohn. After three years of evaluating hundreds of patents under Haller's rigorous demands for analysis and presentation, the specificities of patent work had become, for Einstein, a way of life, a form of work, and (as he suggested to Zangger) a precise and austere style of writing.[111]

Along the same lines, Einstein's relatively accessible framing of the problem of space and time would have come as second nature to a patent examiner. According to Swiss patent law (and not only Swiss law), the description of the invention was to be "representable through a model, the unity of the invention defended, the consequences of the patent must be unambiguously laid out and lucidly ordered, so the whole is easily understood by properly certified technicians as well as by specialists."[112] In all his theoretical papers written around 1905, Einstein ended with a series of assertions about what the experimental consequences would be. In the case of the relativity paper, he concluded with the sharp, numbered paragraphs occasionally used in physics papers but standard in the "claims" section required by Swiss statute to appear at the end of every patent.[113] Strikingly, from 1905 Einstein began describing (and occasionally drawing) devices, not only for his little electrostatic *Maschinchen* but also as key components of his theoretical arguments.

Would General Field Marshal von Moltke have appreciated the irony? Sixteen-year-old Einstein, a young man who had abandoned his German citizenship chafing at the "herd mentality" of the militarist Prussian state, had at age twenty-six, completed the aged officer's project in a certain sense.[114] Time had ever more completely become identified with timekeeping, and *Einheitszeit* a means to the technopolitical establishment of procedural, distant simultaneity across the globe. Einstein's clock synchronization system, like its more mundane predecessors, reduced time to procedural synchronicity, tying clocks together by electromagnetic signals. Indeed, Einstein's scheme for clock unity went much farther, extending beyond city, country, empire, continent, indeed beyond the world, to the infinite, now pseudo-Cartesian universe as a whole.

Here the irony inverts. For while Einstein's clock coordination procedure built on decades of intense efforts toward electromagnetic time unification, he had removed the crucial element of von . Moltke's vision. There was, in Einstein's infinite, imagined clock machine, no national or regional *Primäre Normaluhr*, no *horloge-mère*, no master clock. His was a coordinated system of infinite spatiotemporal extent, and its infinity was without center—no Schlesischer Bahnhof linked upward through the Berlin Observatory to the heavens and down along the rails to the edges of empire. By infinitely extending a time unity that had originally been conceived according to the imperatives of German national unity, Einstein had both completed and subverted the project. He had opened the "zone of unification," but in the process not only removed Berlin as the *Zeitzentrum* but also designed a machine that upended the very category of metaphysical centrality.

Absolute time was dead. With time coordination now defined

only by the exchange of electromagnetic signals, Einstein could finish his description of the electromagnetic theory of moving bodies without spatial or temporal reference to any specially picked-out rest frame, whether in the ether or on earth. No center remained, not even the vestigial centrality of the special ether rest frame that Poincaré had retained. Einstein had constructed his abstract relativity machine out of a material world of synchronized clocks.

Chapter 6

THE PLACE OF TIME

Without Mechanics

B Y DECEMBER 1907, Patent Officer Einstein was no longer an obscure bureaucrat. Minkowski wrote to request a relativity reprint, congratulating the young scientist on his success. Wilhelm Wien debated him on the possibility of faster-than-light signaling. Max Planck and Max Laue joined the young physicist in conversation; one of the best experimentalists in Germany, Johannes Stark, commissioned an article from him on the relativity principle. No longer could Einstein breezily omit the work of his contemporaries; his engagement in the circle of physicists was now active, and his footnotes reflected it. Theorists Emil Cohn and H. A. Lorentz joined Einstein's reference list in his 1907 review of relativity, as did experimentalists Alfred Bucherer, Walter Kaufmann, Albert Michelson, and Edward Williams Morley.[1] Einstein even addressed Lorentz's "local time" directly, as he had not in 1905. But Poincaré's name was nowhere among the thirty-two footnotes. Einstein continued to pass over the older scientist in utter and unbroken silence.[2]

Einstein began by *assuming* that there was no measurable difference between physical processes observed when they were at rest and those same physical processes were they conducted in a constantly moving boxcar. He took as his starting point what Poincaré, Lorentz, and other leading physicists had, earlier in the game, been working painstakingly to *prove*. Poincaré and others had asked *how* electrons flattened as they moved through an all-pervasive ether, *how* the electron could remain stable despite its distortion, *how* the ether reacted as electrified bodies and light passed through it. But in Einstein's paper, all that the French polymath had done vanished. Not just the bits on the ether and electron structure: absent as well was any reference to Poincaré's simplification and correction of the transformations in Lorentz's theory, Poincaré's powerful mathematical advances (including the introduction of the four-dimensional spacetime), Poincaré's articulation of a principled physics, and perhaps most dramatically, Poincaré's interpretation of Lorentz's "local time" as a convention of clock coordination executed by the exchange of light signals. Not a trace.

From Paris, Poincaré echoed back Einstein's silence. Einstein was no doubt a young unknown to him in 1905; there is really no need to explain the absence of any reference to Einstein in Poincaré's 1906 treatise, "On the Dynamics of the Electron." But later, although he and Einstein separately published often on questions of time, space, and the relativity principle, Poincaré's silence continued for seven more years. With Einstein's name on everyone's lips — including Lorentz's, Minkowski's, Laue's, and Planck's — this surely is not happenstance. To Einstein, Poincaré must have seemed beside the point: another older physicist who in 1905 was unable to grasp the import of Einstein's banishment of the ether or his placement of time, without making the true/apparent distinction, at the starting point of the theory. To Poincaré, Einstein must have seemed derivative, perhaps a provider of heuristic arguments to derive the Lorentz transformations, but one who failed even to address fundamental issues of physics: the ether and the structure of the electron.

Yet just as Einstein was in no deep sense derivative, Poincaré could not be dismissed as a mere conservative. On the contrary Poincaré proudly hailed the electrodynamics of moving bodies as a new mechanics when he spoke at Göttingen in 1909; even in St. Louis in 1904 he had heralded dramatic changes throughout physics. At the same time, Poincaré's 1909 presentation at Lille con-

veyed a certain wistfulness as he invoked a classical physics whose
marble columns had begun to crack: "If some part of science
appears solidly established, it would certainly be Newtonian
mechanics; we lean on it with confidence, and it does not seem that
it could ever be weakened. But scientific theories are like empires,
and if Bossuet was here, he would no doubt find eloquent accents
in which to denounce their fragility."[3] Jacques-Bénigne Bossuet,
teacher of the Dauphin, son of Louis XIV, had written for the royal
charge his 1681 *Discourse on Universal History*. Long central within
the canon of French literature, Bossuet's *History* occupied a featured place in a 1912 volume that Poincaré assembled with colleagues from the Académie Française. Aimed at educating the
public in matters both scientific and literary, the volume excerpted
just that part of Bossuet's message that cautioned a prideful humanity: "So you see pass before your eyes, as if in an instant, I say not
kings and emperors but those grand empires that had made the
whole universe tremble? When you see the Assyrians, ancient and
modern, the Medes, the Persians, the Greeks and the Romans
appear before you one after the other and fall, one on top of the
other, so to speak, this terrifying fracas makes you feel that there is
nothing solid among men and that inconstancy and agitation are
the proper fate of things human."[4] For Poincaré, scientific theories
were like Bossuet's grand empires. For decades he had mined the
confrontation among empires of time, crossing the conventions that
moderated the collision with a new philosophy and physics. Yet as
he peered over the two great structures to which he had devoted so
much of his life—the Newtonian empire and the French one—
Poincaré now saw them through the eyes of a man who had registered the fragility of both.

Einstein and Poincaré finally met—for the first and last time—
at the Solvay Conference held in Brussels late in 1911. They were
two scientists infinitely close and yet infinitely distant. Both had
been fascinated since youth by the problem of the electrodynamics

of moving bodies. Both had productively built their theories at the crossroads of technology, philosophy, and physics. Both understood the immense power of Lorentz's work and both underscored the group structure of Lorentz's transformations. Both seized upon the principle of relativity as a constitutive foundation of physics. Perhaps most dramatically, both insisted that time in moving frames had to be interpreted by way of clocks synchronized by light-signal exchange. But the distance between the two scientists was as dramatic as their proximity. Poincaré viewed his contributions as a form of world repair, an adjustment, a turning, a rewriting of Lorentzian physics into a new mechanics that he viewed with exhilaration and trepidation. For the young Einstein, repair held little appeal. Tearing down the old was a bracing pleasure. While Poincaré maintained the ether as crucial in his 1909 Lille address, Einstein began a talk of his own at almost exactly the same time with a specific reference to a physicist (not Poincaré) who had assessed the ether's existence to "border on certainty." Then Einstein knocked the author's assertion into the trash.[5]

For many reasons it is not surprising that the encounter between the fifty-seven year old and the thirty-two year old did not go well. Maurice de Broglie: "I remember one day at Brussels, while Einstein was explaining his ideas, Poincaré asked him, 'what mechanics are you using in your reasoning?' Einstein answered: 'No mechanics' which appeared to surprise his interlocutor."[6] "Surprise" is perhaps an understatement. For Poincaré, whose conception of physics came down to mechanics, be it old or new, "no mechanics" was an impossible response. Abstract mechanics, after all, was what Poincaré had held up to his fellow Polytechnicians as constituting the very essence, the "factory stamp" of that unique training for the world of Third Republic France.

Einstein spoke to his Solvay audience about light quanta and quantum discontinuities, following which the extraordinarily distinguished audience began to debate. Lorentz was there; so was

Poincaré. Einstein initiated the discussion by reminding his listeners that the theory of quanta in its present form really wasn't a theory at all "in the ordinary sense of the word," but a useful tool. Einstein surely did not see his comments as a detailed mathematical description of the type Poincaré would dignify with the term "mechanics." (Einstein had called his account of light quanta a "heuristic" when he had first published it six years earlier.) Rather, Einstein hoped to offer a starting point for a later, more coherent treatment. In the meantime, he had willingly sacrificed the continuous, causal relations that differential equations captured. But that coolness toward the foundation of an intuitive, mathematically describable mechanics was for Poincaré no small matter. Summarizing the conference from his vantage point, Poincaré spoke without minced words. One hears in his remarks a deeply troubled assessment of where the new physics, and the new physicists, were heading:

> What the new research seems to throw into question is not only the fundamental principles of mechanics, it is something that appeared to us till now inseparable from the very concept of a natural law. Could we still express these laws in the form of differential equations?
>
> Besides, what struck me in the discussions that we just heard, is seeing the same theory sometimes relying on the old mechanics and sometimes on new hypotheses that negate them; one must not forget that there isn't a proposition that one can't easily prove insofar as one inserts into the demonstration two contradictory premises.[7]

Here is the pent-up frustration of a master of the "old mechanics," the consternation of someone who had used differential equations to explore, extend, test, and (against his own intent) upend the stability and visualizability of Newtonian physics. It is the voice of a *savant* who has pushed toward a "new mechanics" in every way he knew. Similarly, Poincaré had significantly advanced the Lorentz theory, "turning it in every direction" as he had said so often. In the process, bit by bit, he had grappled for years with changed notions of mass, length, and most dramatically, time itself. More than anyone, he insisted both in his mathematics and in his philosophy that one could switch, according to circumstance, from Euclidian to non-Euclidean language. Shortly after Solvay, Poincaré even advanced understanding of the new and unsettling quantum discontinuity.

In Poincaré we face anyone but a change-abhoring conservative. But he contended that there were better and worse ways of handling innovation. The new physics, Poincaré was saying, had lost its way by abandoning, in its frantic race ahead, the consistent and principled base of *any* — of *all* — mechanics. The "honest functions" and the differential equations that vouchsafed causality and intuition had been lost. This was not a dispute over which law best fit the phenomena. It was a gulf that, for Poincaré, left Einstein and his supporters on the wrong side of "the very concept of natural law." Borrowing Poincaré's own earlier phrase, Einstein's quantum physics seemed less well suited for science than for a teratological museum.

Einstein's reaction to the Solvay encounter with Poincaré was swift and unflattering. A few weeks after Solvay, he confided his views to a friend: "H. A. Lorentz is a marvel of intelligence and tact. He is a living work of art! In my opinion he was the most intelligent among the theoreticians present. Poincaré was simply negative in general, and, all his acumen notwithstanding, he showed little grasp of the situation."[8] There was certainly no meeting of minds between them over the relativity theory. Their split over the quantum widened the chasm.

Yet Poincaré returned to Paris from Solvay in 1911 deeply impressed by Einstein. That November, Einstein, having only recently moved to Prague, was a candidate for a position at his alma mater, the Swiss Federal Institute of Technology. Setting aside any disquiet at Einstein's disturbing radicality, Poincaré intervened in his favor, assuring physicist Pierre Weiss that Einstein was "one of the most original minds I have known." Youth mattered little; the senior scientist judged Einstein to have "already taken a very honorable rank among the leading scholars of his time. What we must above all admire in him, is the facility with which he has adapted to new conceptions and from which he knows how to draw the consequences. He does not remain attached to classical principles, and, in the presence of a problem of physics, is prompt to envision all the possibilities. This translates immediately in his mind into the prediction of new phenomena, susceptible of being one day verified by experiment. I do not want to say that all these predictions will remain impervious to the judgment of experiment when that judgment becomes possible. As he searches in all directions, one must, on the contrary, expect that the majority of the paths in which he embarks will be dead ends; but one must, at the same time, hope that one of the directions that he has indicated will be the right one; and that suffices. It is just so that one must proceed. The role of mathematical physics is to pose questions properly, it is only experiment that can resolve them." This was a letter of the highest praise. "The future will show more and more the value of Mr. Einstein," Poincaré concluded, "and the university that finds a way to secure this young master is assured of drawing from it great honor."[9]

Beyond this statesmanly letter, it is impossible to say precisely what effect Poincaré's one meeting with Einstein had on the senior physicist. Poincaré's health was deteriorating, and he may at this time of enormous productivity also have had intimations of his own mortality. It may also be that Poincaré's later reflections upon Einstein's jarring new vision of physics at Solvay prompted the mathematician to think further on the value of the provisional, heuristic, result-oriented efforts that Einstein had employed to such effect.

Just a few weeks after recommending Einstein to Weiss, on 11 December 1911, Poincaré wrote to the founding editor of *Circolo matematico* of Palermo. Still working on the three-body problem with which he had launched his career decades earlier, Poincaré reported to his correspondent that for two long years he had been struggling with the problem without much progress. He now had to pause, at least temporarily. "It would be fine if I could be sure of being able to take it up again; at my age, I cannot vouch for that, and the results obtained, liable to put researchers on a new and unexplored track, seem to me too full of promises, despite the disappointments that they have caused me, for me to resign myself to sacrificing them." At fifty-seven, Poincaré was hardly old, but just a few years before, he had needed major prostate surgery. Would the editor be willing to publish an incomplete work, one that would state the problem and report partial results? (He would.) "What embarrasses me is that I will be obliged to put in a lot of figures, precisely because I could not arrive at a general rule, but I only accumulated particular solutions." As Poincaré had so often insisted, visual-geometrical intuition could go where skeletal algebra could not yet tread.

Poincaré judged these particular solutions to his lifelong problem "useful." They were more than that; the paper contributed foundational ideas to the establishment of topology, a new branch of mathematics. Soon a young American mathematician, George D. Birkhoff, proved the crucial conjecture that lay at the core of Poincaré's exploration.[10]

Perhaps tacit echoes of Einstein may be audible in the last speech Poincaré gave on relativity, though the presentation never mentioned Einstein's name. On 4 May 1912, Poincaré spoke on "Space and Time" to an audience at the University of London. In forceful terms, he once again repeated: "The properties of time are therefore merely those of our clocks just as the properties of space are merely those of the measuring instruments."[11] Over the past years, the ether had grown ever thinner in Poincaré's writings, as its role in the theory dwindled. Now, however, the all-pervasive substance simply evaporated into silence. No rejection—but no mention of it, either. In his peroration, Poincaré let the old mechanical "principle of relativity" fall away, replaced by the "principle of relativity according to Lorentz." Events simultaneous according to clocks coordinated in one frame of reference would not be simultaneous if measured by clocks coordinated in another.

Does this mean that Poincaré had abandoned the ether or become a thoroughgoing Einsteinian? No. Having begun his presentation by asking if his earlier conclusions about space and time now needed to be revised in light of recent developments, he replied: "Certainly not; we had adopted a convention because it seemed convenient and we had said nothing could constrain us to abandon it." But conventions are not God-given.

> Today some physicists want to adopt a new convention. It is not that they are constrained to do so; they consider this new convention more convenient; that is all. And those who are not of this opinion can legitimately retain the old one in order not to disturb their old habits. I believe, just between us, that this is what they shall do for a long time to come.[12]

Poincaré's time to come was short. His medical difficulties came more frequently and more severely. Nonetheless, when asked to accept the presidency of The French League of Moral Education and deliver its founding address on 26 June 1912, Poincaré characteristically accepted. For him, scientific prestige was inextricably associated with civic leadership and responsibility. In the midst of battles between anticlerical and clerical movements, in the face of an escalating conflict with the Germans in North Africa, even as the walls of Paris were plastered with partisan appeals, Poincaré sought principles that would undergird a unifying French morality. Against those ready to manipulate hatreds, Poincaré saw discipline as the only defense. Discipline—morality—was all that secured mankind against "an abyss of sufferings." "Mankind is . . . like an army at war," an army that must prepare for battle in peacetime, not at the last, too-late moment of engagement with the enemy. Hatred could propel collisions among men, collisions that risked changing their faith. "What will happen if the new ideas which they adopt are those which their former teachers conveyed to them as the very negation of morality? Can this mental habit be lost in one day? . . . Too old to acquire a new education, they shall lose the fruits of the old!"[13] In morality, in physics, in mathematics, Poincaré wanted to build dramatic new structures, but he wanted to do so using the old bricks; he would employ, not discard, the legacy of an illustrious past.

Poincaré underwent surgery again on 9 July 1912, and for a few days friends and family hoped for a recovery. It never came; Poincaré died, following an embolism, on 17 July 1912. Dozens of éloges appeared across the world. Perhaps the most fitting monument was the most anonymous: later that year the Eiffel Tower began radiating its precision time signal. Those pulses bathed the world in an expanding sphere of Hertzian light, fixing simultaneity (and longitude) into Africa and across the Atlantic to North America on the basis of techniques that Poincaré had introduced into geodesy, epistemology, and physics.

Two Modernisms

Reflecting back on Henri Poincaré in 1954, Prince Louis de Broglie, the physicist who had shown that particles could act like waves, lamented that the great mathematician had just missed being the first to develop the theory of relativity in all its generality, "thus gaining for France the glory of that discovery." "It is impossible to be closer to the thought of Einstein," de Broglie judged. "And yet

Poincaré did not take the decisive step; he left to Einstein the glory of grasping all the consequences of the principle of relativity and, in particular, of establishing, by a profound criticism of the measure of lengths and durations, the true physical character of the relation that the principle of relativity has between space and time. Why did Poincaré not come to the end of his thought? It is no doubt the turn, a little too critical, of his spirit, due perhaps to his education as pure mathematician. . . ." For de Broglie, it was Poincaré's training as a mathematician that had led him to see science as no more than the informed and expeditious choice, on the basis of convenience, of one theory from among all logically equivalent ones. According to de Broglie, Poincaré failed to tread the better path laid down by the physicist's intuition.[14] By de Broglie's lights, Poincaré was too much a mathematician, too indifferent to the real world to have formulated relativity as Einstein did.

My view? De Broglie's diagnosis is far too narrow. I would argue that Poincaré did come to the "end of his thought," to an image of knowledge — including his view of mathematical knowledge — that carried with it a nineteenth-century optimism, a Third Republic Polytechnician's engaged, hopeful vision of a calculable, improvable, rational world. If anything, Poincaré paid too much attention to the real world: when he judged in 1898–99 that the corrections to Newtonian time were in principle necessary but too small to matter, it was because at that moment he was assessing the light-signal "relativistic" errors against the real-world "ordinary" longitude-timing errors. Yet to call Poincaré's approach "conservative" or "reactionary" is to miss the point; Poincaré's sight-line aimed directly toward the ideals of a revolutionary Enlightenment that, by century's end (Poincaré's time), had grown into institutionalized French empire. All our great constructions eventually crack, Poincaré says on many occasions. But our response to these fissures, these crises, should not be the mysticism or the melancholy of the intellectual elites, but instead a redoubled effort to repair those breaks by the systematic application of reasoned action. As Poincaré saw it, the scientist-engineer could apply analytical reason as readily to the understanding of a coal-mine accident as to planetary motion, as easily to the mapping of the world as to the reconstruction of Lorentz's theory of moving electrons.

For Poincaré, the trunk of the tree of knowledge was precisely this engaged mechanics, an intuitively grounded mathematical understanding of nature that, at myriad points, shot branches into experiment and technology. Searching in a thousand ways, Poincaré aimed for an understanding of the world that could on the one hand speak to students trying to comprehend their place in an injured France and on the other hand to scientists, cartographers, and politicians struggling to wire together an empire that would bind Paris to Dakar, Haiphong, and Montreal. He wanted a mechanics of forces and energy, but one sufficient to undergird analysis of celestial mechanics, the shape of the earth, or the behavior of telegraph wires. As he reminded his fellow *anciens polytechniciens* in 1903, the required mixture was one of theory and action. In Poincaré's case that meant at one time responding to British telegraphic hegemony by fostering French radio signaling, at another by dismantling the unscientific indictment against Dreyfus with astronomical instruments and the calculus of probabilities.

His was a world where truth and the ultimate reality of things meant far less than the establishment of communicable, stable, durable relations—the kind of reliable relations that made action possible. As Poincaré put it, "science is only a classification and . . . a classification can not be true, merely convenient. But it is true that it is convenient, it is true that it is so not only for me, but for all men; it is true that it will remain convenient for our descendants; it is true finally that this can not be by chance. In sum, the sole objective reality consists in the relations of things."[15] A world of scientific rationality without metaphysical profundity: objective relations, not metaphysical objects.

For Poincaré, joining the abstract and concrete in this flat world of relations meant being able to negotiate hard-fought conventions in the human world. Sorting and negotiating the needs and demands of railroad magnates, astronomers, physicists, and navigators lay front and center in the decimalization of time. And as a leading figure at the Bureau of Longitude, Poincaré grasped time through the detailed, material procedures of engineering protocols: organizing, analyzing, reporting on the expeditions of the military-scientific colleagues he so admired as they hammered together observatory shacks in the high Andes or on the coastal stations of Senegal. Time, for Poincaré, resided in *our* world, *our* convenience, *our* exchange of optical and electrical telegraph signals. The metaphysical world behind appearances was nothing. As he wrote in "The Measure of Time," "We . . . choose these rules [of simultaneity] not because they are true but because they are the most convenient."

For Poincaré the choice of how to measure simultaneity made time richer, not poorer. It meant he could work the time concept back and forth among the protocols of longitude or decimalization, the abstractions of a science-inflected philosophy, and the principles of a new physics. "Let us watch [scientists (*savants*)] at work and look for the rules by which they investigate simultaneity," Poincaré urged.[16] This is precisely what Poincaré did as he struggled to synthesize the work of his circle of philosophers, physicists, and cartographers. Time-as-procedure stood in all three series:

> *Simultaneity is a convention, nothing more than the coordination of clocks by a crossed exchange of electromagnetic signals taking into account the transit time of the signal.*

This move has provided the central drama of this book, charging its historical moment with a critical opalescence. What was it? In one sense it was a conventional, regulated procedure, a practical, ever-more-precise method for the day-to-day establishment of simultaneity for the sake of fixing longitude, a theory machine. As a technical convention, it occurred again and again in the pages of the *Annals of the Bureau of Longitude*. At the same time, for Poincaré it was a philosophical exploration of questions of time and simultaneity, a statement that he could present as his prime example of a *conventional* stance toward scientific laws and principles. Simultaneity was no more than electromagnetic coordination grounded in principled agreement. In this philosophical register, the utterance appeared appropriately and dramatically in the *Review of Metaphysics and Morals*, fitting perfectly into the longer conversation Poincaré had been conducting with a circle of French philosophers, many of whom had emerged from Polytechnique. Finally, beginning in 1900, Poincaré presented this simple simultaneity statement to an audience of physicists as an interpretation of Lorentz's "local time" as if Lorentz had implicitly and all along held such a view. When Poincaré turned to theorizing about the electron, the simultaneity procedure could grace the published proceedings of a physics conference dedicated to Lorentz or the *Proceedings of the Academy of Sciences*.

Is clock coordination "really" a technological, a metaphysical, or a physical intervention? All three. We may as well ask if the Place de l'Etoile is truly in the Avenue des Champs Elysées, the Avenue Kleber, or the Avenue Foch. In fact, as in a great metropolitan intersection, the enormous interest of the simultaneity question lies precisely in its position at the center of a vibrant crossing of great intellectual avenues.

Repeated ceaselessly from East Africa to the Far East, the procedure of electromagnetic clock coordination was at once thoroughly technological and altogether theoretical. It was brass tubes with fragile, suspended mirrors, and it was global control of "universal time." It was vast lengths of copper cable protected by heavy gutta-percha insulation, lying a mile under the ocean and served by brass telegraph keys inside crude observatory shacks; it was also the phantasmagorical reach of empire. Surveyors and astronomers pored over personal equations, calculating and correcting measurements through simultaneity on their way to the map, the final, prized product of their longitude work. But clock coordination was also the assembly of synchronized clocks strung together like beads along a continent-spanning chain through Europe, Russia, and North America. This fusion of technology and science drew together (though often in all-out struggle) Swiss *horlogiers*, American train schedulers, British astronomers, and members of the German General Staff. At one extreme, time synchronization represents technology conditioning humdrum daily procedure at every two-bit whistle-stop. As such it is a substantial part of the thick social history of the New England or Brandenburg countryside. At the other extreme, it stands for the symbolic reach of modernity that had mayors, physicists, and philosophers, pronouncing on the conventionality of time while poets payed homage to the annihilation of space by speed. In this register, clock coordination was rarified history, pursued at the pinnacle of European philosophy and mathematical physics.

For Poincaré, the modern technology of time was not external to his scientific life — not a "context" that from some mythical "outside" shaped, influenced, or distorted thought. Poincaré was in and of this compound world, a product of and professor in the Ecole Polytechnique where the material and the abstract shaped one another at every moment. He proudly bore the "factory stamp." Poincaré's work on time was of a historical era and of a place; one aspect is not an accidental influence on another. To externalize Ecole Polytechnique or the Bureau of Longitude as entities outside Poincaré's true self is to break linkages among technical and cultural actions that he and his late-nineteenth-century contemporaries saw as joined. Not only had Poincaré held the presidency of the Bureau on three separate occasions, but he also served as one of its elite Academy members for twenty years, published regularly in its journal, and played leading roles in its most active committees on the measurement of time. Nor is it "external" to Einstein's physics that he trained at ETH, an institution committed by its very charter to joining theory and praxis, or that he capped that instruction with a seven-year apprenticeship at the Bern patent office, as a quality control officer in the production-machine of modern electrotechnology. No, these are not influences moving Einstein and Poincaré from the outside. They were rather fields of action that conveyed the high value of machines grasped through reason—production sites for technology (through science) and science (through technology).

Circa 1900, to speak of the transmission-corrected synchronization of clocks was both central and ordinary. Transmission correction of time was a working tool for the longitude finder and routine for city engineers in Paris and Vienna, who knew about time delay all too well through their frustrating attempts to pump exact time through pneumatic tubes under their cities. By 1898, the transmission delay of a time pulse was a standard problem for the squadrons of engineers, cartographers, physicists, and astronomers who were creating simultaneity every day of the week.

So in January 1898, when Poincaré published his argument for time-as-convention, he was speaking the technology of clocks as well as the language of philosophy. The same utterances now could be heard in different registers. "Simultaneity is a convention," or "synchronizing clocks demands transmission-corrected electrical coordination"—Are these philosophical observations or do they "really" belong to brass and copper technology? Is the Place de l'Etoile really in the Avenue Kleber or the Avenue Foch?

In December 1900, Poincaré cleared a third avenue through the Place de la Simultanéité, when he began using clock synchronization—first approximately and later, exactly—to assign meaning to Lorentz's local time. At that point three fields were fully engaged: telegraphic longitude, philosophical conventionalism, electrodynamic relativity. It is our loss that we have dessicated this remarkable moment, split it into fragments, and scattered them over the disconnected academic departments of philosophy, physics, and metrology. Poincaré struggled to hold that modern and modernizing world together—to fix it, uphold it, defend it.

Young Einstein also stood in the midst of this trading zone of philosophy, technology, and physics. But he was never out to repair and uphold any empire—neither the French, nor the Prussian, nor the Newtonian. Delightedly mocking senior physicists, teachers, parents, elders, and authority figures of all kinds, happily calling himself a "heretic," proud of his dissenting approach to physics, Einstein shed the nineteenth century's ether with an outsider's iconoclastic pleasure. Not for him was the increasingly desperate hunt for a stable basis of the Solar System, or for a bedrock foundationalism that would ground all of physics in mechanics or electrodynamics. Instead, Einstein was content—more than content—to find theory-devices that worked. Heuristics, temporary but effective means of going forward with the theory, were machines. So was his light clock, or the myriad of machinelike thought experiments that he proposed for thinking through the inertia of energy, $E = mc^2$. And so, most importantly for our purposes, was his time machine, that infinite array of clocks connected and coordinated by the well-regulated exchange of light signals.

As important as time coordination was for Poincaré as a pragmatic, conventional aid to the building of a new mechanics, it was more crucial for Einstein. For Einstein it was procedurally defined time that served as the starting point from which the Lorentz contraction would be derived. Like a classical arch, for Einstein time synchronization held the principled column of relativity in stable union with the principled column of the absolute speed of light.

Did Einstein really discover relativity? Did Poincaré already have it? These old questions have grown as tedious as they are fruitless. No doubt originally propelled by the offensive and widespread

Nazi-era assaults on Einstein's place in physics, the struggle over "who discovered relativity" continued for decades: Who found the theory? What is its essence? Is the core of relativity *really* the dismissal of the ether, or is it the mathematical formula for transforming space and time? Is it an unshakeable commitment to the relativity principle or is it the applicability of the principle to all physical interactions or is the theory rightly identified with the derivation of time and space transformations from the synchronization of clocks? Or is relativity really no more than correct predictions of what can be observed in experiments? Above all, relativity—and the relativity of time in particular—became synonymous with modern physics and modernity more generally. From our perspective, assigning a graded checklist to Einstein and Poincaré counts as the least interesting part of the story of time and simultaneity.

Far more important is to situate Poincaré and Einstein at the two nodal points of turn-of-the-century time coordination, grasping the characteristic ways in which each navigated the flows of technology, physics, and philosophy, and understanding how each struggled to rip simultaneity from the metaphysical firmament and bring it to earth as a procedurally defined quantity. Time standardization was the order of the day, for each scientist a natural extension of the standardization of length. It announced itself on the faces of public clocks, railroad schedules, and inside regulated schoolrooms and factory floors. In the Paris Observatory of the 1890s, Wolf was winding electromagnets to keep astronomical clocks in step and supervising the distribution of dozens more throughout the streets of Paris. Cornu was adjusting the giant pendulum on his regulator clock, joining mechanics and electromagnetism to formulate a rigorous analysis of electrosynchronization. Teams of itinerant observers worked incessant time-exchanges with Senegal, Quito, Boston, Berlin, and Greenwich. Anglo-Saxon astronomers hawked time to railroads, while French stargazers hoped the glorious precision of the observatory would be imitated through the whole of the country—reflections of a mother clock cast from mirror to mirror until the light of temporal rationality spilled through every street in the republic.

Creating this standardized, procedural time was a monumental project that utilized creosote-soaked poles and undersea cables. It required a technology of metal and rubber, but also reams of paper, bearing, contesting, and sanctifying local ordinances, national laws, and international conventions. As a result, conventional, turn-of-the-century time synchronization never inhabited a place isolated from industrial policy, scientific lobbying, or political advocacy. It would ease matters if we could attribute the late-nineteenth-century push toward standards to a single drive-wheel: if we could say that it all came down ultimately to railroad magnates or decisively to scientists or exclusively to philosophers. But the restructuring of time was not that simple.

Time was complex because, long before the nineteenth century, clocks and simultaneity were already more than gears, pendula, and pointers. In the eighteenth century, for example, the precision chronometers of the long-suffering British clockmaker John Harrison appeared in a larger print world about the longitude problem as well as in diaries and satires about timekeeping and mapping. From the start, Harrison's precision clocks assumed a "planetary" significance.[17] Forging back earlier than the eighteenth century still does not shake timepieces loose from the cultures in which they were built. Sand clocks and church clocks carried much besides the assignment of time; they conveyed different, overlapping authorities of God, of the feudal lord, of the memory of mortality. There simply is no getting behind the cultural to some primordial moment in which time was nothing but sand, shadow, or a mechanical pointer.

What emerged in the late nineteenth century was not merely the issue of a particular invention—coordinated clocks certainly existed before then. Instead, Europe and North America experienced a dramatic, global alteration in the place and density of electrodistributed time. Linked clocks of the late nineteenth century covered the world. Circles of such wide-spanning technologies pulled each other along. Trains dragged telegraph lines, telegraphs made maps, maps guided rail-laying. All three (trains, telegraphs, maps) contributed to a growing sense that long-distance simultaneity made the question, What time is it now somewhere else? at once practical and evocative. When we read Einstein's and Poincaré's many discussions of the new ideas of time and space cast in the lexicon of automobiles, telegraphs, trains, and cannons, we are seeing the conditions under which these questions became commonplace.

We can think of simultaneity as the intersection of arcs, where each arc stands for a long sequence of "moves" within a field. Take the physics we have followed. Clearly there is no single, unchanged meaning of simultaneity from the early 1890s through 1905. Local time (*Ortszeit*) began in a geographical sense, became Lorentz's fictionally offset local time, turned into Poincaré's light-signal *observable* offset local time, shifted to Poincaré's offset and dilated "apparent" time, only to take a new form in Einstein's relativistic time. These shifts in the meanings of time did not take place all at once and did not play out purely in the domain of physics. Instead, they are better understood as a series of moves in an evolving game. Consistent with the use of "move" in the everyday sense of a game, a "move" in this more technical sense was sometimes a statement (or convention), sometimes a physical procedure. Remarkably, there was enough sense of continuity for Poincaré and Lorentz to see their work as building step by step, even though the aims of the game they were playing were also evolving. (If in 1894 Lorentz's goal was to solve an equation by making a moving object in an electric and magnetic field look as if it were at rest in the ether, his goal [and Poincaré's] in 1904–05 was to produce laws of physics that led to the same measurable results in any constantly moving frame of reference.)

One arc of simultaneity, then, was that of physics—the series of moves transforming the electrodynamics of moving bodies. But Poincaré was playing his light-signal synchrony move in at least two other arcs as well: telegraphic longitude determination and late-nineteenth-century French philosophy. From the time of the American Civil War forward, telegraphic longitude became *the* modern method for determining simultaneity for longitude. Pushed by the Coast Survey, Europeans rapidly took up the American method, deploying it over land and under sea. By 1899, under Poincaré's presidency, the Bureau of Longitude had become a global node for sending, receiving, processing, and defining simultaneity. Time at the end of the nineteenth century was written in devices all through the Bureau: methods for distributing time to cities, theories of how to improve electrosynchronization, debates over the decimalization of time. Electric longitude was important for the French to fix their colonies on the worldmap and articulate their internal geographies. But it was also crucial to counter the insult to European geodesy presented by the longstanding battle over the right longitudinal difference between Paris and London. Poincaré in the lead, the Bureau enlisted the Eiffel Tower in wireless time transmission. Inscribed in report after report from the Bureau, the exchange of telegraphic signals using transmission times became ordinary work in the fixing of time and place far from Paris. Philosophy too had its arc, its series of statements about the measure of time. No doubt Poincaré could see philosophy through the work of the Boutroux circle, even more proximately through the physics-philosophy of his fellow Polytechnicians Auguste Calinon and Jules Andrade, who in their own ways were dissecting time.

The intersection of these three registers of time measurement do not *necessitate* Poincaré's revision of simultaneity. But probing the triple intersection of concerns at sites like Polytechnique and the

Bureau of Longitude does offer a recognition of why Poincaré then and there would seize the measurement of time as an essential question tied to conventions, physics, and longitude. Of why abstract time could be grasped and reconceived through machines.

Each of the three arcs—of physics, philosophy, and technology—carried with it a sense of the new. The "new mechanics" advertised its rupture with old notions of mass, space, and time, the electric world-spanning telegraph cables was a celebrated triumph and tool of "civilizing" empire. Poincaré linked his conventionalism about time and simultaneity with his philosophical conventionalism about the principles of physics and the fabric of mathematics. But conventions for Poincaré could also refer to the plethora of French-based international conventions about the measure and distribution of the precise hour. Time stood un-still at this quintessentially modern triple intersection.

Throughout his work, Poincaré treated his subjects with a mathematical engineer's modernism: that is, with a deep-seated faith in the human ability to grasp and improve the world technically, to map it, so to speak, all the way down. Just after Poincaré's death, his nephew, Pierre Boutroux, struggled in a letter to Mittag-Leffler to capture the animating goal of his uncle's life work. Intriguingly, he turned not to mathematics or physics, but rather to geography for his guiding thread. Boutroux recounted how all his life Poincaré had avidly followed stories of exploration and travel; both inside and outside science, all his work was characterized by a drive "to fill the white spaces on the map of the world."[18]

It was Poincaré's abiding faith that the blank spaces on the worldmap could be filled. Gaps on the surface of knowledge could be completed in the causal map Poincaré drew of the Magny mining disaster, tracking the catastrophe all the way back to the pickax dent in the latticework of mining lantern 476. He would track those lacuna more globally through the Poincaré Map, which he exploited to chart the unvisualizable behavior of chaotic planetary orbits as successions of points meandered over the plane. To fill in blanks on the geographical worldmap, he worked with the Bureau of Longitude to plot telegraphic maps of Saint-Louis, Dakar, and Quito, work that also had implications for theoretical attempts he and a long tradition of physicists had made to account for the shape of the earth. Other white spaces might be replaced by studies of the ether, by investigating the structure of the electron, by insistence on intuitive mathematical functions, intuitive formulations of logic, and the productive intuitions afforded by the ether.

Poincaré's was a hopeful modernism of relations graspable by us, without God, without Platonic forms, and (though he was fascinated by Kant's emphasis on structures through which experience becomes possible) without Kantian things-in-themselves. Instead of attending to objects, Poincaré was forever after relations, for it was the relation of things that would survive even when the objects that they tied together had faded behind the mists of history. Truth? Given the complexities of conventions, definitions, and principles that went into the laws of physics, he preferred the objectivity that came with shared, durable concepts of simplicity, and convenience. True relations, not truth by itself. Visible surfaces, not obscure depths. Poincaré pursued this reformulated Enlightenment vision even if it meant driving radically new concepts of space, time, and physical stability into the vast blank spaces of knowledge.

Einstein's modernism too can be found at a triple intersection: a move in the physics of moving bodies, a move in the philosophical attack on absolute time and space, and a move in the wider technology of clock synchronization. Einstein's focus was more insistently physical than Poincaré's, more attendant to particular material machines rather than to the engineering of abstract ones. It is impossible to imagine Poincaré spinning an ebonite wheel in a homebuilt contraption; it is equally hard to picture Einstein coordinating a massive team effort to engineer the cabling of precision electric time from Quito to Gayaquil. While maintaining a more hands-on engagement with the materiality of objects, Einstein also held a more metaphysical conception of the relation of theories to phenomena, one that led him in many different contexts to demand a sharp correspondence between elements of the theory and elements of the world. Poincaré never gave up his assignment of local time to the status of "apparent" in contrast to "true." Because he saw no parallel distinction in the phenomena themselves, Einstein wouldn't touch such a theoretical dichotomy. Time and space in one inertial frame were as "true" (or "relative") for Einstein as in any other: clocks were clocks, rulers were rulers. Where Poincaré kept the ether as a means-for-thinking, an intuitive basis on which differential equations could be imagined, Einstein mocked the ether as a remnant idle gear from an obsolete mechanism. And he threw it aside with the same relish that he mustered for patent applications with superfluous elements. When Einstein handled the light quantum heuristically, without reference to the wave equations understood within the ether, it left Poincaré fearing that the young physicist and his supporters had jettisoned the very conditions that made possible real understanding of the physical world. In his terms, Poincaré was right: Einstein was perfectly willing to use intellectual devices as stop-gaps—heuristics that tied elements of theory to elements of the phenomena, even if that meant violating intuition (in Poincaré's particular sense).

Poincaré struggled to map the world through differential equations, chosen for convenience in the largest sense, all the way to tertiary rivulets feeding secondary streams. Did the ether and "apparent time" aid intuition while preserving the "true relations" of observed phenomena? Then for Poincaré that was satisfactory, even if it meant a certain redundancy. By contrast, Einstein wanted to orient time and space within a theory that *matched* the phenomena, not just in prediction but in austerity. If the phenomena were symmetric (for example, if there were no way to distinguish the moving magnet/still coil from still magnet/moving coil), then the theory should formally encapsulate that symmetry. Later, in the quantum debate, Einstein expressed the complementary concern: that there were predictable features of the physical world to which no element of the theory corresponded.

For Poincaré space and time were pinned to the rigorous surface of objective relations built to meet our human need for a frankly psychological, objective, and simple convenience. His was a relentless Third Republic secularism. By contrast, Einstein did not think that theory had fulfilled its task by successfully and conveniently capturing true relations among phenomena. He aimed for a depth between phenomena and the theory that underlay them. Like Poincaré, Einstein believed that laws must be simple, not for our convenience but because (as Einstein put it) "nature *is* the realization of the simplest conceivable mathematical ideas." The form of the theory therefore had to exhibit in its detailed form the reality of the phenomena: "In a certain sense," Einstein later insisted, "I hold it true that pure thought can grasp reality, as the ancients dreamed."[19] Einstein believed that a proper theory would match the phenomena in austerity. In that depth lay a contemplative theology. Not the religiosity of a personal, vengeful, or judgmental God, but a mostly hidden God of an underlying natural order: "The scientist is possessed by the sense of universal causation. The future to him is every whit as necessary and determined as the past. . . . His religious feeling takes the form of a rapturous amazement at the harmony of natural law which reveals an intelligence of such superiority."[20] Sometimes it was given to the physicist to advance by the provisional application of heuristic devices; these could tide the theory over until further development was possible. Such a provisional use of formal principles played a role in thermodynamics, in quantum theory, and in relativity.[21] But Einstein insisted over and over that, insofar as they could, scientists fashioned theories that seized some bit of the underlying, simple, and harmonious natural order. Since

Einstein believed that the phenomena did not distinguish true from apparent time, neither, he insisted, should the theory.

Neither Poincaré nor Einstein falls into a naive realism or anti-realism. True, throughout his career, Poincaré underscored the freedom of choice present in describing the world: in geometry, in physics, in technology. But it would thoroughly misrepresent his position to lump his conventionalism with an anything-goes anti-realism. Both in practical and abstract matters he took every opportunity to emphasize the central role of objective, "true relations," of a simplicity that was not up for grabs. Einstein, by contrast, is frequently pigeonholed as straightforwardly realist; there is, after all, the Einstein who comfortably asked if a theory was "the true Jacob." Nonetheless, he cautioned that there are different ways to characterize "reality" and that the fecund part of theory lay not in the events that were classifiable with spacetime coordinates, nor even with the directly perceptible. Instead, it lay in the links between them, and these links were not fixed once and for all.[22] Both scientists recognized the power of principles and conventions in shaping the reach of theory and the possibility of measurement. Both were fully prepared to reject concepts even if those received concepts seemed to carry along history, intuitiveness, and self-evidence. Deeply embedded in a changing electrotechnical world that more than at any time previously recognized the importance of *choice* in measurement, standardization, and theory construction, Einstein and Poincaré separately cracked simultaneity from its metaphysical pedestal and replaced it by a convention given through machines.

Reading *back* from Einstein it is all too easy to cast Poincaré as a reactionary, striving toward (but failing to reach) Einstein's theory of relativity. Such a retrospective view would bury Poincaré's reworking of the physics of space and time into a "new mechanics." It would be as if Picasso were to be jettisoned as antimodern because he was not modern in the sense of Pollock, or to do the same to Proust because his was not the modernism of late Joyce.

Reading *forward* to both Poincaré and Einstein, we can see each breaking with the past in different ways.

Here were two great modernisms of physics, two ferociously ambitious attempts to grasp the world in its totality. Poincaré's modernism advanced by establishing objective, simple, convenient, and true relations down to the smallest white space. Einstein's moved forward by chiseling out a theory that aspired to capture the phenomena, not just in predictions, but in its underlying structures. One was constructive, building up to a complexity that would capture the structural relations of the world. The other was more critical, more willing to set aside complexity in order to grasp, austerely, those principles that reflected the governing natural order. These twin visions of a new and modern relativistic physics had much in common. Yet Einstein and Poincaré remained in ambivalent admiration of each other, no more able to engage each other's alternate modernity than Freud could read Nietzsche. Too close and too far to speak, their skew interpretations of relativity never crossed, even as the two scientists radically altered "time" in ways that shook knowledge in physics, philosophy, and technology.

From a biographical point of view, it is of course remarkable that Einstein and Poincaré were able to participate in such variegated technical and philosophical activities as if they were chess grand masters, playing simultaneous championship games and finding a *single* successful move that checkmated them all. Yet chess provides only a weak analogy. These "games" of physics, philosophy, and technology were of vastly different construction, the consequences of the simultaneity move enormous in each domain. Philosophers of the Vienna Circle, no less than leading physicists in the 1920s, and engineers of the Global Positioning System of the 1980s all looked to Poincaré-Einstein simultaneity as a model for the construction of future scientific concepts.

In the end, however, the opalescent history of time coordination is falsely presented by reducing it to biography. Cropping the picture to portrait size renders invisible the vast, disputed, standardized conventions of measurable time and space that coursed through Europe and the United States. This is not because Einstein's or Poincaré's imaginations were too limited but instead because "the measure of time" fluctuated across so many scales. Time coordination had become a modern problem through the conventionality and regulation of time flowing through observatories, cables, train networks, and cities. A better analogy is this: Isobars and isotherms transformed and in part made possible predictive meteorology. Similarly, the electric world array of clocks made distant synchronization into a quintessentially modern problem resolvable by a machinelike procedure—even if that machine turned out to be both infinite and theoretical.

Times and places where the technical, philosophical, and scientific are all *centrally* implicated are rare, much less frequent even than the kind of physics developments traditionally described as "revolutions." In the nineteenth century entropy and energy may offer similarly scale-shifting histories: think of the fateful intersection of steam engines, thermodynamics, and quasi-theological discussions about the inexorable "heat-death" of the Universe. To find a more recent mixture of abstraction and concreteness of this kind, we can look to the mid-twentieth-century explosion of "information sciences": cybernetics, computer science, cognitive science. Here converged the dense histories of wartime feedback devices that tumbled out of weapons production, alongside the more arcane trajectories of information theory and models of the human mind. Time, thermodynamics, computation: each defined an age symbolically and materially. Each represented a moment of critical opalescence when it became impossible to think abstractly without invoking machines or to think materially without grasping for world-spanning concepts.

Looking Up, Looking Down

Times changed. Einstein left the Bern patent office on 15 October 1909 for the University of Zurich; on 1 April 1911, he began his tenure at the Karl-Ferdinand University in Prague and, in the spring of 1914, joined the University of Berlin. There he both completed his general theory of relativity and became a leading spokesman against the war. After the fighting had ended, Favarger, that avatar of Swiss chronometric unity whom we met earlier, published his 550-page third edition of his technical treatise on electrical time-keeping, framing its detailed electromechanical content, once again, in broadly cultural terms. The Great War, he argued, had contributed powerful technological developments, but it had also destroyed a great part of the human wealth that sustained peace had created. What remained was by contrast "a heap of ruins, miseries and suffering."[23] Humanity needed work to overcome this disaster, and work invariably involved time.

Time, Favarger rhapsodized, "cannot be defined in substance; it is, metaphysically speaking, as mysterious as matter and space." (Even stolid Swiss clockmakers apparently were driven to metaphysics by time.) All the activities of man, conscious or unconscious, sleeping, eating, meditating, or playing take place in time. Without order, without specified plans, we risk falling into the anarchy Favarger had warned against since long before Gavrilo Principe shot Archduke Ferdinand. Now, after the Great War, the risk loomed larger that people could fall into "physical, intellectual and moral misery." His remedy? the precise measurement and determination of time with the rigor of an astronomical observatory. But to function as an antidote, measured time could not remain in the astronomers' redoubt; time rigor must be distributed electrically to anyone who wanted or needed it: "we must, in a word, popularize it, we must *democratize* time" in order for people to live and prosper. We must make every man "*maître du temps,* master not only of the hour but also of the minute, the second, and even in special cases the tenth, the hundredth, the thousandth, the millionth of a second."[24] Distributed, coordinated precision time was more than money for Favarger, it was each person's access to orderliness, interior and exterior—to freedom from time anarchy.

Throughout the late nineteenth and early twentieth centuries, coordinated clocks were never just gears and magnets. Certainly time was more than merely technical for Poincaré and Einstein. It was also much more than wires and escapements for New England village elders, time-zone campaigners, Prussian generals, French metrologists, British astronomers, and Canadian promoters. In the opalescent history of time coordination, clocks trapped nerve transmissions and reaction times, structured workplaces and guided astronomy. But the two great scale-changing domains of material time centered on the railroad and the map. The Bureau of Longitude that Poincaré had helped supervise stood as one of the great time centers of the world for the construction of maps. And the Swiss Patent Office, where Einstein had stood guard as a patent sentinel, was the great Swiss inspection point for the country's technologies concocted to synchronize time in railways and cities.

My hope in exploring clock coordination has been to set Poincaré's and Einstein's place within a universe of actions that crossed mechanisms and metaphysics, that made abstract concreteness or, if you will, concrete abstractions. More generally, perhaps we can begin to look at science in a way that avoids two equally problematic positions on the relation of things to thoughts. On one side, there is a long tradition of what could be described as a reductive materialism, the view that demotes ideas, symbols, and values to surface ripples on a deeper current of objects. Through those empiricist glasses of the 1920s through the 1950s, theoretical physics and its philosophy often seemed a provisional addition, not the bulwark of science. Einstein (on this view) appeared as having taken the last, inexorable step in an inductive process that gradually drove out the ether and the absolutes of space and time. Earth's motion through the ether could not be detected to an accuracy better than the ratio of earth's velocity to the speed of light (v/c, that is, about one part in ten thousand). Later such measurements improved, showing no evidence of motion to a much higher accuracy (second order in v/c, or about one part in a hundred million), and *therefore*, so the argument went, Einstein concluded that the ether was "superfluous."[25] No doubt there is much to be said for this experiment-grounded Einstein. His fascination with the detailed conduct of experiments and his gyrocompass work at the Physikalisch-Technische Reichsanstalt reveal a theorist with a clear sense of laboratory procedure and the operation of machines. On the empiricist view, things structured thoughts.

On the flip side was the antipositivist movement popular in the 1960s and 1970s. Thoughts structured things. Antipositivists aimed to reverse the older generation's epistemic order; they saw programmes, paradigms, and conceptual schemes as coming first, and they held these to have completely reshaped experiments and instruments. Einstein on the antipositivist screen appeared as the philosophical innovator who dispensed with the material world altogether in a sustained drive for symmetry, principles, and operational definitions. There is much truth here too—reading history with antipositivist glasses exposes those moments when Einstein was chary of experimental results, dubious, for example, of supposed laboratory refutations of special relativity and of astronomical observations that claimed to threaten the general theory.

Granting both ways of reading history their due, I do not mean to split the difference. Instead, attending to moments of critical opalescence offers a way out of this endless oscillation between thinking of history as ultimately about ideas or fundamentally about material objects. Clocks, maps, telegraphs, steam engines, computers all raise questions that refuse a sterile either/or dichotomy of things *versus* thoughts.[26] In each instance, problems of physics, philosophy, and technology cross. Staring through the metaphorical we can find the literal; through the literal we can see the metaphorical.

When Einstein came to the Bern patent office in 1902, he entered an institution in which the triumph of the electrical over the mechanical was already symbolically wired to dreams of modernity. Here clock coordination was a practical problem (trains, troops, and telegraphs) demanding workable, patentable solutions in exactly his area of greatest professional concern: precision electromechanical instrumentation. The patent office was anything but the lonely deep-sea lightboat that the no longer young Einstein had longed for as he spoke to the Albert Hall audience in the dark days of October 1933. Reviewing one patent drawing after another in the Bern office, Einstein had a grandstand seat for the great march of modern technologies. And as coordinated clocks were paraded by, they were not traveling alone. The network of electrical chrono-coordination provided political, cultural, and technical unity all at once. Einstein seized on this new, conventional, world-spanning simultaneity machine and installed it at the principled beginning of his new physics. In a certain sense he completed the grand time coordination project of the nineteenth century by designing a new, vastly more general time machine valid everywhere and for all imaginable constantly moving frames of reference within the Universe. But by eliminating the master clock and redefining conventionally defined time to a *starting point*, Einstein came to be seen by both physicists and the public as having changed their world.

Poincaré, at the end of his life, co-authored a book entitled *What Books Say, What Things Say*. This quirky volume combined the endeavors of the two great academies to which he belonged: the literary Académie Française and the scientific Académie des Sciences. From the literary side came articles on the heros of culture, including Hugo, Voltaire, and Bossuet. Poincaré himself contributed chapters on stars, gravity, and heat, but also on coal mining, batteries, and dynamos. Living as easily among philosophers as among mathematicians and engineers, Poincaré's scholarship, including his work on simultaneity, stood central to all these cultures.

Einstein's stance toward the great academies of science was different. Just after World War I, Arthur Eddington, the British astrophysicist, took advantage of a total eclipse to measure the deflection of starlight by the gravitational pull of the sun (or as Einstein would have it, by the sun's bending of spacetime). Thrust into front-page fame by the resulting confirmation of his general theory of relativity, Einstein overnight became a world figure. Facing the limelight in his increasingly public role, his relativity work after 1919 moved toward the abstract unification of physical forces and away from the machine culture of the patent office. A few days after his 1933 speech in Royal Albert Hall, Einstein left for the United States, where he embraced a venerated if monastic existence at the Institute for Advanced Study in Princeton. Half seer, half mascot, he spoke in oracular terms on everything from the meaning of God to the future of nuclear warfare. In April 1953, two years before his death, Einstein wrote from Princeton to Maurice Solovine of the laughter and insight of their so-unacademic academy during Einstein's Bern years, when patents, physics, and philosophy stood side by side:

To the immortal Olympia academy,

In your short active existence you took a childish delight in all that was clear and reasonable. Your members created you to amuse themselves at the expense of your big sisters who were older and puffed up with pride. I learned fully to appreciate just how far [the members had through you] hit upon the true through careful observations lasting for many long years.

We three members, all of us at least remained steadfast. Though somewhat decrepit, we still follow the solitary path of our life by your pure and inspiring light; for you did not grow old and shapeless along with your members like a plant that goes to seed. To you I swear fidelity and devotion until my last learned breath! From one who hereafter will be only a corresponding member, A.E.[27]

As each struggled with time, philosophy, and relativity at the turn of the century, Poincaré inhabited the Parisian académies of arts and sciences, Einstein the Olympia (non-) Academy. On 15 March 1955, Michele Besso died. It was with Besso that Einstein had so productively spoken in the weeks and months before he came upon the coordination of clocks as the key to finishing his work on special relativity. Einstein wrote to Besso's family on the twenty-first, closing his letter with a final reference to those conversations, and to the perspectival nature of time that emerged from relativity: ". . . it was the Patent Office that reunited us [after Zurich]. Our conversations on the way home had an incomparable charm; it was if the all-too-human did not exist at all. . . . Now he has also taken leave of this strange world a little before me. This means nothing. For us devout physicists, the division between past, present, and future is only an illusion, if a stubborn one."[28]

Long after Einstein's own last learned breath, the struggle continued among many competing interpretations of the regulated coordination of clocks. Synchronized time remained hypersymbolized. *Einheitszeit* never emerged from contestation among imperial empire, democracy, world citizenship, and antianarchism. What all these symbols held in common was a sense that each clock signified the individual, so that clock coordination came to stand in for a logic of linkage among people and peoples that was always flickering between the literal and the metaphorical. Precisely because it was abstract-concrete (or concretely abstract), the project of time coordination for towns, regions, countries, and eventually the globe became one of the defining structures of modernity. The synchrony of clocks remains an inextricable mix of social history, cultural history, and intellectual history; technics, philosophy, physics.

Over the last thirty years it has become a commonplace to pit bottom-up against top-down explanations. Neither will do in accounting for time. A medieval saying aimed at capturing the links between alchemy and astronomy put it this way: In looking down, we see up; in looking up, we see down. That vision of knowledge serves us well. For in looking down (to the electromagnetically regulated clock networks), we see up: to images of empire, metaphysics, and civil society. In looking up (to the philosophy of Einstein and Poincaré's procedural concepts of time, space, and simultaneity) we see down: to the wires, gears, and pulses passing through the Bern patent office and the Paris Bureau of Longitude. We find metaphysics in machines, and machines in metaphysics. Modernity, just in time.

NOTES

CHAPTER 1

1. Einstein, "Autobiographical Notes" [1949], 31. On the universal "tick-tock" see Einstein, "The Principal Ideas of the Theory of Relativity" [after December 1916], *Collected Papers*, vol. 7, 1–7, on 5. On Newtonian time and space: Rynasiewicz, "Newton's Scholium" (1995).

2. We can now read Einstein's work through the extraordinary scholarship of several generations of historians. This literature is so vast that I can refer only a few sources here—they serve as entry points into the wider literature: for both superb editorial comments and meticulous reproduction of documents, Stachel et al., eds., *Collected Papers* (1987–); for secondary literature, Holton, *Thematic Origins of Scientific Thought* (1973); Miller, "Einstein's Special Theory of Relativity" (1981); Miller, *Frontiers* (1986); Darrigol, *Electrodynamics* (2000); Pais, *Subtle is the Lord* (1982); Warwick, "Role of the Fitzgerald-Lorentz Contraction Hypothesis" (1991); idem, "Cambridge Mathematics and Cavendish Physics" (part I, 1992; part II, 1993); Paty, *Einstein philosophe* (1993); M. Janssen, *A Comparison between Lorentz's Ether Theory and Special Relativity in the Light of the Experiments of Trouton and Noble*, unpublished doctoral dissertation, University of Pittsburgh, 1995; and Fölsing, *Albert Einstein* (1997). For collections of essays by leading scholars, see "Einstein in Context," *Science in Context* 6 (1993), and Galison, Gordin, and Kaiser, *Science and Society* (2001). For an extensive bibliography of other historical works on special relativity: Cassidy, "Understanding" (2001).

3. The scholarship on Poincaré, also vast, is now coming into its own with the work of the Nancy-based project of the Archives Henri Poincaré, which is publishing the scientific correspondence. See, for example, Nabonnand, ed., *Poincaré–Mittag-Leffler* (1999); published articles are mostly in *Oeuvres* (1934–53). An overview of current work on Poincaré's technical work may be found in the literature of note 2 above (especially works by Darrigol and Miller) along with references cited there, as well as the volume by Paty on the links between Poincaré's physics and philosophy; see also the excellent volume, Greffe, Heinzmann, and Lorenz, eds., *Henri Poincaré, Science and Philosophy* (1996). Rollet's excellent dissertation

surveys Poincaré's role as a popularizer and philosopher—it also contains a fine bibliography; see Henri Poincaré, "Des Mathématiques à la Philosophie. Études du parcours intellectuel, social et politique d'un mathématicien au début du siècle," unpublished doctoral dissertation, University of Nancy 2, 1999.

4. Galison, "Minkowski's Space-Time" (1979).

5. Einstein, "Elektrodynamik bewegter Körper" (1905), 893; I have used a (slightly modified) version of the translation given in Miller, Einstein's Special Theory of Relativity (1981), 392–93.

6. Ibid.

7. See the sources in note 2 above; on the ether, Cantor and Hodge, eds., Conceptions of Ether (1981).

8. For Heisenberg on his discussions with Einstein about the critique of absolute time, Physics and Beyond (1971), 63; other quantum theorists (Max Born and Pascual Jordan) also modeled their new physics on Einstein's simultaneity convention, Cassidy, Uncertainty (1992), 198; Philipp Franck reported the "good joke" remark in Einstein (1953), 216.

9. Schlick, "Meaning and Verification" (1987), 131; see also 47.

10. Quine, "Lectures on Carnap," 64.

11. Einstein, Einstein on Peace (1960), 238–39, on 238.

12. Einstein, Autobiographical Notes [1949], 33.

13. Barthes, Mythologies (1972), 75–77.

14. Poincaré, "Mathematical Creation" [1913], 387–88.

15. Poincaré, Science and Hypothesis (1952), 78.

16. Quoted in Seelig, ed., Helle Zeit-dunkle Zeit (1956), 71; trans. in Calaprice, The Quotable Einstein (1996), 182.

17. Remotely set clocks were discussed by, among others, Charles Wheatstone and William Cook, the Scottish clockmaker, Alexander Bain, and the American inventor Samuel F. B. Morse. For Wheatstone, Cooke, and Morse, clock coordination came out of their work on telegraphy. See Welch, Time Measurement (1972), 71–72.

18. For pre-1900 discussions of the extensive work on clock coordination, see, for example, the series of articles by Favarger, "L'Electricité et ses applications à la chronométrie" (Sept. 1884–June 1885), esp. 153–58, and "Les Horloges électriques" (1917); Ambronn, Handbuch der Astronomischen Instrumentenkunde (1899), esp. vol. 1, 183–87. On the expansion of the Bern network, see the Gesellschaft für elektrische Uhren in Bern, Jahresberichte, 1890–1910, Stadtarchiv Bern.

19. Bernstein, Naturwissenschaftliche Volksbücher (1897), 62–64, 100–104. I would like to thank Jürgen Renn for helpful discussions about Bernstein.

20. Poincaré, "Measure of Time" [1913], 233–34.

21. Ibid., 235.

22. Poincaré, "La Mesure du temps" (1970), 54. Slightly modified.

Chapter 2

1. Poincaré, "Les Polytechniciens" (1910), 266–67.

2. Ibid., 268, 272–73.

3. Ibid., 274–75, 278–79.

4. Cahan, *An Institute for an Empire* (1989), esp. ch. 1.

5. Monge's polestar, descriptive geometry, plummeted in curricular importance from 153 hours (in 1800) to 92 hours (in 1842). Meanwhile, analysis, the rigorous study of mathematical functions, climbed to the top from its secondary role. Belhoste, Dahan, Dalmedico, and Picon, *La formation polytechnicienne* (1994), 20–21; Shinn, *Savoir scientifique et pouvoir social* (1980). On Monge's projective geometry, see Daston, "Physicalist Tradition" (1986). On pedagogy in physics more generally, see the Warwick articles cited above and Olesko, *Physics as a Calling* (1991), and David Kaiser, *Making Theory: Producing Physics and Physicists in Postwar America*, unpublished doctoral dissertation, Harvard University, 2000.

6. Poincaré on Cornu, "Cornu" (1910), esp. 106, 120–21, originally published in April 1902 (cf. Laurent Rollet, *Henri Poincaré. Des Mathématiques à la Philosophie. Étude du parcours intellectuel, social et politique d'un mathématicien au début du siècle*, unpublished doctoral dissertation, University of Nancy 2, 1999, 409); Cornu, "La Synchronisation électromagnétique" (1894).

7. Picon opposes this distant respect for experiment to Terry Shinn's characterization of the curriculum as more frankly hostile to experiment. Belhoste, Dahan, Dalmedico, and Picon, *La formation polytechnicienne* (1994), 170–71; Shinn, "Progress and Paradoxes" (1989).

8. See Poincaré to his mother, e.g., C76/A74, C97/A131, C112/A150, C114/A152, C116/A162, in *Correspondance de Henri Poincaré* (unpubl. Archives—Centre d'Études et de Recherche Henri Poincaré, 2001), all from the academic year 1873–74.

9. C79/A92, in *Correspondance de Henri Poincaré* (unpubl. Archives—Centre d'Études et de Recherche Henri Poincaré, 2001).

10. Roy and Dugas, "Henri Poincaré" (1954), 8.

11. See Nye, "Boutroux Circle" (1979). Quotation from Archives—Centre d'Études et de Recherche Henri Poincaré microfilm 3, n.d. (probably 1877) in Laurent Rollet, *Henri Poincaré. Des Mathématiques à la Philosophie. Étude du parcours intellectuel, social et politique d'un mathématicien au début du siècle*, unpublished doctoral dissertation, University of Nancy 2, 1999, 78–79, on 79; also cf. 104. For the limits of science debate, see Keith Anderton, *The Limits of Science: A Social, Political and Moral Agenda for Epistemology*, unpublished doctoral dissertation, Harvard University, 1993.

12. Calinon, "Étude Critique" (1885), 87.

13. Ibid., 88–89; letter Calinon to Poincaré, 15 August 1886, in *Correspondance de Henri Poincaré* (unpubl. Archives—Centre d'Études et de Recherche Henri Poincaré, 2001).

14. Calinon to Poincaré, 15 August 1886, in *Correspondance de Henri Poincaré* (unpubl. Archives—Centre d'Études et de Recherche Henri Poincaré, 2001).

15. Roy and Dugas, "Henri Poincaré" (1954), 20.

16. Ibid., 18.

17. Ibid., 17–18.

18. Ibid., 23; on Caen appointment see Gray and Walter, *Henri Poincaré* (1997), 1.

19. On Poincaré's emphasis on curves see Gray, "Poincaré" (1992); Gilain, "La théorie qualitative de Poincaré" (1991); Goroff, Introduction, in Poincaré, *New Methods* (1993), I9. Two fine articles on Poincaré and chaos are Gray, "Poincaré in the Archives" (1997), and (more technical) Andersson, "Poincaré's Discovery of Homoclinic Points" (1994).

20. Goroff, Introduction, in Poincaré, *New Methods* (1993), I9, from Poincaré, "Mémoire sur les courbes" [1881], 376–77. Emphasis added.

21. Poincaré, "Sur les courbes définies par les équations différentielles" [1885], 90; cited and translated in Barrow-Green, "Poincaré" (1997), 34.

22. Barrow-Green, "Poincaré" (1997), 51–59.

23. Poincaré, "La Logique et l'intuition" [1889], 132.

24. Mittag-Leffler to Poincaré, 16 July 1889, letter 89, in *La Correspondance entre Poincaré et Mittag-Leffler* (1999).

25. "I had thought that *all* these asymptotic curves, having moved away from a closed curve representing a periodic solution, would then asymptotically approach the *same* curve." Poincaré to Mittag-Leffler, 1 December 1889, letter 90, in *La Correspondance entre Poincaré et Mittag-Leffler* (1999).

26. Poincaré to Mittag-Leffler, 1 December 1889, letter 90, in *La Correspondance entre Poincaré et Mittag-Leffler* (1999).

27. Mittag-Leffler to Poincaré, and attached notes, 4 December 1889, letter 92, in *La Correspondance entre Poincaré et Mittag-Leffler* (1999).

28. Weierstrass to Mittag-Leffler, 8 March 1890, note to letter 92, in *La Correspondance entre Poincaré et Mittag-Leffler* (1999).

29. For an accessible discussion of chaotic phenomena, see Ekeland, *Mathematics* (1988), and Diacu and Holmes, *Celestial Encounters* (1996), from which the following figures are drawn; more technical discussions in Barrow-Green, "Poincaré" (1997); and Goroff, Introduction, in Poincaré, *New Methods* (1993).

30. Poincaré, *New Methods* (1993), part 3, section 397, 1059.

31. On the postmodern interpretation of chaos, see, for example, Hayles, *Chaos and Order* (1991); Wise and Brock, "The Culture of Quantum Chaos" (1998); for

physics and the connection of chaotic physics and art, see Eric J. Heller, www.ericjhellergallery.com (accessed June 19, 2002).

32. Poincaré, "Sur le problème des trois corps" [1890], 490; Poincaré, Preface to the French Edition [1892], in idem, *New Methods* (1993), xxiv.

33. Poincaré, Preface to the French Edition [1892], in idem, *New Methods* (1993), xxiv.

34. Poincaré, "Sur les hypothèses fondamentales" [1887], 91.

35. Poincaré, "Non-Euclidean Geometries" [1891], in idem, *Science and Hypothesis* (1902), 50, 41–43.

36. Giedymin, *Science and Convention* (1982), 21–23, on 23.

37. On Riemann, for example, cf. A. Gruenbaum, "Carnap's Views" (1963); and his *Geometry and Chronometry* (1968). On Helmholtz as a source for Poincaré, Gerhard Heinzmann, "Foundations of Geometry," *Science in Context* 14 (2001), 457–70. On reading sources of Poincaré's mathematical conventionalism from more contemporary sources such as Jordan or Hermite, see Gray and Walter, Introduction, in *Henri Poincaré* (1997), 20.

38. Poincaré, "Non-Euclidean Geometries" [1891], in idem, *Science and Hypothesis* (1902), on 50, emphasis added.

CHAPTER 3

1. Duc Louis Decazes, in *Documents diplomatiques* (1875), see 36. For an excellent introduction to the joint moral and technical history of standardization, see Wise, ed., *Precision* (1995), and further references in the rich essays by Simon Schaffer, M. Norton Wise, Graham Gooday, Ken Alder, Andrew Warwick, Frederic Holmes, and Kathryn Olesko. On the original mission to fix the meter see Alder, *Measure* (2002).

2. J.B.A. Dumas, in *Documents diplomatiques* (1875), 121–30, esp. 126–27.

3. Guillaume, "Travaux du Bureau International des Poids et Mesures" (1890).

4. Comptes rendus des séances de la première conférence générale des poids et mesures, Poincaré, "Rapport" (1897). After the burial of **M**, metrological work moved toward the adoption of a different sort of procedure, one in which the wavelength of light became the standard of length—and so replacing that particular bar with a certain number of cesium wavelengths. This amalgamation of metrologists and spectroscopists, astrophysicists, and optical physicists is tracked by two fine pieces: Charlotte Bigg, *Behind the Lines. Spectroscopic Enterprises in Early Twentieth Century Europe*, unpublished doctoral dissertation, esp. Part II, University of Cambridge, 2002; and Staley, "Traveling Light" (2002).

5. "Le Nouvel étalon du mètre" (1876).

6. *Le Temps*, 28 September 1889, 1.

7. See, for example, in *Comptes rendus de l'Académie des Sciences*: Violle, "Sur l'alliage du kilogramme" (1889); Larce, "Sur l'extension du système métrique" (1889); Bosscha, "Études relatives à la comparaison du mètre international" (1891); Foerster, "Remarques sur le prototype" (1891), 414.

8. "Extrait du Rapport du Chef du Service Technique," Ponts et Chaussées, 5 March 1881. Archives de la Ville de Paris, VONC 219.

9. Dohrn-van Rossum, *History of the Hour* (1996), 272.

10. "Conseil de l'Observatoire de Paris, Présidence de M. Le Verrier" [1875]; Le Verrier to M. le Préfet [January?, 1875?], both Archives de la Ville de Paris, VONC 219.

11. "Projet d'Unification de l'heure dans Paris. Rapport de la Commission des horloges," 22 January 1879. Archives de la Ville de Paris, VONC 219.

12. Tresca, "Sur le réglage électrique de l'heure" (1880); Ingénieur en Chef, Adjoint aux Travaux de Paris, "Quelques Observations en Réponse au Rapport du 25 Novembre, 1880." Archives de la Ville de Paris, VONC 3184, 6.

13. See, for example, G. Collin to M. Williot, 23 September 1882; G. Collin to M. Chrétien, 10 April 1883; both Archives de la Ville de Paris, VONC 219.

14. Breguet, "L'unification de l'heure" (1880); on time coordination in Paris, see also David Aubin, "Fading Star," forthcoming.

15. M. Faye to M. le Directeur, Direction des Travaux de Paris, 16 January 1889. Archives de la Ville de Paris, VONC 219.

16. Nordling, "L'Unification" (1888), 193.

17. Ibid., 198, 200–201, and 202.

18. Ibid., 211.

19. Sobel, *Longitude* (1995), and Bennett, "Mr. Harrison" (2002).

20. G. P. Bond to A. D. Bache, Supt USCS. 28 Feb. 1854. Harvard University Archives, Harvard College Observatory, Cambridge, MA. Chronometric Expedition, letters, reports, miscellany; Box 1: Reports.

21. W. C. Bond, "Report of the Director" (4 December 1850).

22. W. C. Bond, "Report of the Director" (4 December 1851), clvi-clvii; G. P. Bond to A. D. Bache, Supt. USCS. 22 Oct. 1851. Both documents in Harvard University Archives, Harvard College Observatory, Cambridge, MA. Chronometric Expedition, letters, reports, miscellany; Box 1: Reports.

23. Stephens, "Partners in Time" (1987), 378.

24. Stephens, "'Reliable Time'" (1989), 17; who cites, on 19, Shaw, *Railroad Accidents* (1978), 31–33.

25. Bartky, *Selling Time* (2000), 64.

26. "Report of the Director" (1853), clxxi.

27. The literature on individual observatories is simply immense, and there is

no better survey of their various roles in time coordination than Bartky, *Selling Time* (2000), which concentrates on the American case.

28. Jones and Boyd, "The First Four Directorships" (1971), on 160; agreement with Boston & Providence Railroad, Boston & Lowell, Eastern Railroad Company, Boston and Maine Co., etc. Harvard University Archives, Harvard College Observatory, Cambridge, MA. Observatory Time Service, 1877–92. Box 1, folder 7.

29. "Historical Account" (1877), 22–23.

30. Pickering, in idem, *Annual Report of the Director* (1877), 10–11.

31. Report from Leonard Waldo, assistant to Prof. Edward C. Pickering, Director of the Observatory Harvard College, 20 November 1877, Appendix C in Pickering, *Annual Report* (1877), 28–36.

32. George H. Clark, Proprietor, Rhode Island Card Board to Director of Cambridge Observatory, 16 May 1877. Harvard University Archives, Harvard College Observatory, Cambridge, MA. Correspondence re: Time Signals. Folder 2.

33. Waldo, "Appendix C" (1877), 28–29.

34. Ibid., 33–34.

35. Charles Teske to Leonard Waldo, 12 July, 8 August, 15 August, and 11 November 1878. Harvard University Archives, Harvard College Observatory, Cambridge, MA. Correspondence re: Time Signals. Folder 1.

36. Leonard Waldo, handwritten report to Director for year ending 1 Nov 1878. Harvard University Archives, Harvard College Observatory, Cambridge, MA. Observatory Time Service, 1877–92. Box 1, folder 8.

37. Charles Teske to Leonard Waldo, [illegible day] December 1878 and 15 April 1879. Harvard University Archives, Harvard College Observatory, Cambridge, MA. Correspondence re: Time Signals. Folder 1; "Law of Connecticut, approved March 9, 1881," Statutes of Conn., 1881, Ch. XXI. Harvard University Archives, Harvard College Observatory, Cambridge, MA. Observatory Time Service, 1877–92. Box 1, folder 6; S. M. Seldon (Gen. Mgr. New York & New England RR Co.) to W. F. Allen, 23 March 1883, William F. Allen Papers, New York Public Library Archives, New York City, NY. Incoming Correspondence: Box 3, book 1.

38. T. R. Welles to L. Waldo, 5 December 1877. Harvard University Archives, Harvard College Observatory, Cambridge, MA. Correspondence re: Time Signals. Folder 2.

39. *Proceedings of the American Metrological Society* (1878), 37.

40. Bartky, "Adoption of Standard Time" (1989), 34–39.

41. "Report of Committee on Standard Time" (May 1879), 27.

42. Ibid.

43. W. F. Allen to Cleveland Abbe, 13 June 1879. William F. Allen Papers, New York Public Library Archives, New York City, NY. Outgoing Correspondence: Box 3, book VII, 1; the two Time Conventions merged in 1886 becoming American

Railway Association, later Association of American Railroads. See Bartky, "Invention of Railroad Time" (1983), 13.

44. Cleveland Abbe to W. F. Allen. 14 June, 1879. William F. Allen Papers, New York Public Library Archives, New York City, NY. Incoming Correspondence: Box 3, book I.

45. Charles Dowd had in mind a system not far from what was adopted. But when in 1879 he asked Allen if he, Dowd, could prepare an article explaining his idea for "National Time" in Allen's Railway Guide, Allen demurred, saying that there simply was no space for the intervention. Charles F. Dowd to W. F. Allen, 30 October 1879. William F. Allen Papers, New York Public Library Archives, New York City, NY. Incoming Correspondence: Box 3, book I; W. F. Allen to Dowd, 9 December 1879. William F. Allen Papers, New York Public Library Archives, New York City, NY. Outgoing Correspondence: Book VII. Letters reprinted in Dowd, *Charles F. Dowd* (1930), IX.

46. On Fleming, see Blaise, *Time Lord* (2000); Creet, "Sandford Fleming" (1990); older literature includes Burpee, *Sandford Fleming* (1915). Quotation in Fleming, "Terrestrial Time" (1876), 1.

47. Fleming, "Terrestrial Time" (1876), 14–15.

48. Ibid., 31, 22, and 36–37.

49. Fleming, "Longitude" (1879), 53–57; attack on the French position on 63.

50. Cleveland Abbe, U.S. Signal Office, to Sandford Fleming, 10 March 1880. Barnard to Fleming, 18 March, 6 April 1880, and 29 April 1881. All Barnard to Fleming letters Sandford Fleming Papers, National Archives of Canada, Ottawa, Ontario, MG 29 B1 Vol 3. File: Baring-Barnard. For Barnard showing a watch, see *Proceedings of the American Metrological Society* (1883).

51. Barnard to Fleming, 11 June 1881, Sandford Fleming Papers, National Archives of Canada, Ottawa, Ontario, MG 29 B1 Vol 3. File: Baring-Barnard; Smyth, "Report to the Board of Visitors" (1871), R12–R20, on R19; response to Smyth's work more generally by Barnard, "The Metrology" (1884). On Smyth and his natural theological metrology, see Schaffer, *Metrology* (1997).

52. Airy to Barnard, 12 July 1881, Fleming Papers, vol. 3, folder 19.

53. Barnard to Fleming, 19 August, 3 September, and 8 September 1881, quotation from 3 September. Sandford Fleming Papers, National Archives of Canada, Ottawa, Ontario, vol. 3, folder 19. On the shambles of Thomson's appointment: Barnard, "A Uniform System" (1882).

54. John Rodgers to Hazen, 11 June 1881. United States Naval Observatory LS-M vol. 4.

55. "Report of the Committee" [December 1882].

56. Col. H. S. Haines (Gen. Mgr., Charleston & Savannah Railway) to W. F.

Allen, 12 March 1883. William F. Allen Papers, New York Public Library Archives, New York City, NY. Incoming Correspondence: Box 3, book I, 72.

57. Allen, *Report on Standard Time* (1883), 2–6. W. F. Allen, Scrapbook, at Widener Library Harvard University.

58. Allen, *Report on Standard Time* (1883), 5.

59. Ibid., 6.

60. Figures on the number of letters and telegrams from Allen, "History" (1884), 42; figures on the number of lines running from different cities' times from Bartky, "Invention of Railroad Time" (1983), 20.

61. Newspaper article, unsigned, in letter to W. F. Allen from F. C. Nunenmacher (Central Vermont Railroad), 23 November 1883. William F. Allen Papers, New York Public Library Archives, New York City, NY. Incoming Correspondence: Box 5, book IV, 158; E. Richardson (for D. D. Jayne and Son, Publishers, Philadelphia) to W. F. Allen, 5 December 1883. William F. Allen Papers, New York Public Library Archives, New York City, NY. Incoming Correspondence: Box 5, book V, 48.

62. See, for example, telegram of Col. A. A Talmage (Gen. Transport. Mgr. of the Missouri Pacific Railway), in Allen, "History" (1884), 42; S.W. Cummings (Central Vermont Railroad) to W. F. Allen, 26 November 1883. William F. Allen Papers, New York Public Library Archives, New York City, NY. Incoming Correspondence: Box 5, book V, 18; and George Crocker (Asst. Supt., Central Pacific R. R. San Francisco) to W. F. Allen, 8 October 1883. William F. Allen Papers, New York Public Library Archives, New York City, NY. Incoming Correspondence: Box 4, book II, 97.

63. John Adams (Gen. Supt., Fitchburg Railroad) to W. F. Allen, 2 October 1883. William F. Allen Papers, New York Public Library Archives, New York City, NY. Incoming Correspondence: Box 4, book II, 68; W. F. Allen to John Adams, 4 October 1883. William F. Allen Papers, New York Public Library Archives, New York City, NY. Incoming Correspondence: Box 3, book VII, 299.

64. *Proceedings of General Time Convention* (11 October 1883).

65. Ibid.

66. For U.S. and Canada, *Proceedings of the Southern Railway Time Convention* (17 October 1883).

67. *Proceedings of the General Time Convention* (11 October 1883).

68. W. F. Allen to Mayor Franklin Edson requests shift of New York time to 75th meridian; followed by Edson to Alderman, 24 October 1883, in J. S. Allen, ed., *Standard Time* (1951), 17.

69. 7 November 1883, in J. S. Allen, ed., *Standard Time* (1951), 18.

70. Barnard to Fleming, 22 October 1883. Fleming Papers.

71. de Bernardières, "Déterminations télégraphique" (1884). On de Bernardières: *Dossier sur Octave, Marie, Gabriel, Joachim de Bernardières*, November 1886. Archives of the Service historique de la marine, Vincennes, No. 2879.

72. Figure; details from www.porthcurno.org.uk/refLibrary/Construction.html, (accessed 14 February 2002).

73. Green, *Report on Telegraphic Determination* (1877), 9–10.

74. *Report of the Superintendent of Coast Survey* (1861), 23.

75. Before the Civil War, the process had a rather improvisational quality. Dudley Observatory in Albany, New York, for example, aimed to determine its relative position to New York City by lofting a wire from a small wooden building assembled on the observatory site to the private residence of astronomer Lewis M. Rutherfurd, esquire, at Second Avenue and Eleventh Street in New York City. By B. A. Gould with observers George W. Dean with Edward Goodfellow, A. E. Winslow and A. T. Mosman, appendix 18, in *Report of the Superintendent of the Coast Survey* (1862), 221–23.

76. Introduction, in *Report of the Superintendent of the Coast Survey* (1864); also ibid. Gould, appendix 18, 154–56; *Report of the Superintendent of the Coast Survey* (1866), esp. 21–23.

77. *Report of the Superintendent of the Coast Survey* (1867), esp. 1–8 and ibid. Gould, appendix 14, 150–51.

78. *Report of the Superintendent of the Coast Survey* (1867), 60.

79. See, for example, Prescott, *History* (1866), esp. ch. XIV; Finn, "Growing Pains at the Crossroads" (1976); and Provincial Historic Site, "Heart's Content Cable Station" (www.lark.ieee.ca/library/hearts-content/historic/provsite.html, accessed on 8 April 2002).

80. On Gould's earlier American work and his adoption of British techniques, see Bartky, *Selling Time* (2000), 61–72. On the transatlantic work, see Gould, in *Report of the Superintendent of the Coast Survey* (1869), 60–67.

81. Gould, in *Report of the Superintendent of the Coast Survey* (1869), 61.

82. Gould, in *Report of the Superintendent of the Coast Survey* (1869), 63, 65.

83. *Report of the Superintendent of the Coast Survey* (1873), 16–18; appendix 18 in *Report of the Superintendent* (1877), 163–64. Triangle in ibid., 164.

84. Green, *Report on the Telegraphic Determination* (1877).

85. Green, Davis, and Norris, *Telegraphic Determination of Longitudes* (1880).

86. Ibid., 8.

87. Ibid., 9.

88. Davis, Norris, and Laird, *Telegraphic Determination of Longitudes* (1885), 10.

89. Ibid., 9.

90. de Bernardières, "Déterminations télégraphiques" (1884).

91. La Porte, "Détermination de la longitude" (1887).

92. Rayet and Salats, "Détermination de la longitude" (1890), B2.

93. Annex III in *International Conference at Washington* (1884), 210.

94. *Septième Conférence Géodésique Internationale* (1883), 8.

95. *International Conference at Washington* (1884), 24.

96. Ibid., 37.

97. Ibid., 39–41, on 39.

98. Ibid., 41.

99. Ibid., 42–47, on 47.

100. Ibid., 42, 44, and 49–50.

101. Ibid., 51.

102. Ibid., 52–54.

103. Ibid., 54.

104. Ibid., 62–64, on 64.

105. Ibid., 65–68, quotations on 65, 67, 68.

106. Ibid., 68–69.

107. Ibid., 76–80.

108. Ibid., Lefaivre on 91–92; adoption of the metric system on 92–93; Thomson on 94; the vote, see 99.

109. Ibid., 141.

110. Ibid., 159 and 180.

111. On the history of the revolutionary calendar in France, see Baczko, "Le Calendrier républicain" (1992); Ozouf, "Calendrier" (1992).

112. *International Conference at Washington* (1884), 183–88, on 184.

CHAPTER 4

1. Janssen, "Sur le Congrès" (1885), 716.

2. At the Paris Telegraphic Conference of 1890, the assembled urged adoption of a universal time that would be set for all the world by the time of clocks at the prime meridian; cf. *Documents de la Conférence Télégraphique* (1891), 608–9.

3. Howard, *Franco-Prussian War* (1979), quotation on 2, see also 43.

4. Bucholz, *Moltke* (1991), ch. 2 and 3, esp. 146–47; also Bucholz, *Moltke and the German Wars* (2001), 72–73, 110–11, and 162–63.

5. The establishment of uniform time is discussed in Kern, *Culture* (1983), 11–14; and in Howse, *Greenwich* (1980), 119–20. Simon Schaffer uses the Wells time machine as a guide through to the turn-of-the-century intersection of the mechanized workplace, literary and scientific engagement with time in "Time Machines" (in his "Metrology," 1997).

6. Moltke, "Dritte Berathung des Reichshaushaltsetats" (1892), 38–39 and 40;

trans. Sandford Fleming, under the title "General von Moltke on Time Reform" (1891), 25–27.

7. Fleming, "General von Moltke on Time Reform" (1891), 26.

8. Newspaper clippings from the Cambridge University Library, including P.S.L, "Fireworks at the Royal Observatory," *Castle Review* (n.d.); Nigel Hamilton, "Greenwich: Having a Go at Astronomy," *Illustrated London News* (1975); and Philip Taylor, "Propaganda by Deed—the Greenwich Bomb of 1894" (n.d.). Conrad, *Secret Agent* (1953), 28–29.

9. Lallemand, *L'unification internationale des heures* (1897), 5–6.

10. Ibid., 7.

11. Ibid., 8 and 12.

12. Ibid., 17, 18, and 22–23.

13. Poincaré, "Rapport sur la proposition des jours astronomique et civil" [1895].

14. President of the Bureau of Longitude, 15 February 1897, *Décimalisation du temps et de la circonférence*, executing an order of the Minister of Public Instruction, 2 October 1896. From Archives Nationales, Paris.

15. *Commission de décimalisation du temps*, 3 March 1897.

16. Ibid., 3.

17. Ibid., 3.

18. Noblemaire to President Loewy, 6 March 1897; printed in *Commission de décimalisation du temps*, 3 March 1897, 5.

19. Bernardières to Monsieur le Président du Bureau des Longitudes, 1 March 1897, printed in *Commission de décimalisation du temps*, 3 March 1897, 7.

20. Bureau de la Société francaise de Physique to M. le Ministre du Commerce, approved by the Conseil de la Société on 22 April 1897; reproduced in Janet, "Rapport sur les projets de réforme" (1897), 10.

21. In addition to the factor of 4 just mentioned, the division of the circle into 400 parts had a factor of 6 to convert 24 hours of time per day into the 400 grads that divided the circle (dividing 24 by 400 yielded a factor of 6); the final factor of 9 entered when one wanted to convert between old angles and new ones—multiplying by 360 and dividing by 400. The chart itself is reproduced in Poincaré, "Rapport sur les résolutions" (1897), 7.

22. The Sarrauton system is presented in Sarrauton, *Heure décimale* (1897), with Sarrauton's contribution dated April 1896.

23. *Commission de décimalisation du temps*, 7 April 1897, 3.

24. Cornu, "La Décimalisation de l'heure" (1897).

25. Ibid., 390.

26. Poincaré, "La décimalisation de l'heure" [1897], 678 and 679.

27. Note pour Monsieur le Ministre, 29 November 1905, Archives Nationales, Paris, F/17/2921.

28. Sarrauton, *Deux Projets de loi*, addressed to Loewy at the Bureau des Longitudes, 25 April 1899, Observatoire de Paris Archives, 1, 7, and 8.

29. La Grye, Pujazon, and Driencourt, *Différences de longitudes* (1897), A3.

30. Headrick, *Tentacles* (1988), 110–13.

31. La Grye, Pujazon, and Driencourt, *Différences de longitudes* (1897), A6.

32. Ibid., A13, citation on A84.

33. La Grye, Pujazon, and Driencourt, *Différences de longitudes* (1897), A135–36.

34. Headrick, *Tentacles* (1988), 115–16.

35. A discussion of the Paris-London longitude campaign can be found in Christie, *Telegraphic Determinations* (1906), v–viii and 1–8, and further references therein. On the desirability of a redetermination, see the International Geodetic Conference, Paris, 1898.

36. Lectures of 1892–93 reprinted in Poincaré, *Oscillations Electriques* (1894); idem, "Etude de la propagation" [1904], on submarine case, see 454.

37. *Report of the Superintendent of the Coast Survey* (1869), 116.

38. Loewy, Le Clerc, and de Bernardières, "Détermination des différences de longitude" (1882), A26 and A203.

39. Rayet and Salats, "Détermination de la longitude" (1890), B100.

40. La Grye, Pujazon, and Driencourt, *Différences de longitudes* (1897), A134.

41. Calinon, *Étude sur les diverses grandeurs* (1897), 20–21.

42. Calinon, *Étude sur les diverses grandeurs* (1897), 23 and 26. Former Polytechnician Jules Andrade said much the same thing ("there is an infinity of admissible clocks") in his book on the foundations of physics, *Leçons de méchanique physique* (Paris, 1898), 2, which he finished on 4 September 1897. Poincaré cited this work too in "The Measure of Time" to support the contention that when we choose one clock over another it is a matter of convenience, not of one running true and the other wrong. Though Poincaré pursued the quantitative "scientific" notion of silmultaneity while Bergson attended principally to the qualitative experience of time, Bergson's *Time and Free Will* ([1889], 2001) focused attention on the meaning of time.

43. Note pour Monsieur le directeur, 20 March 1900. Archives Nationales, Paris, F/17/13026. Martina Schiavon tracks the role of the army, surveying, and the *savant-officier* in her richly documented study, "*Savants officiers*" (2001). For a textured study of colonialism and surveying (and many further references) see Burnett, *Masters* (2000).

44. *Comptes rendus de l'Association Géodésique Internationale* (1899), 3–12

October 1898, 130–33, 143–44; Poincaré's remark in *Comptes rendus de l'Académie des Sciences* 131 (1900), Monday, 23 July, 218.

45. Headrick, *Tentacles* (1988), 116–17.

46. Poincaré, "Rapport sur le projet de revision" (1900), 219.

47. Ibid., 221–22.

48. Ibid., 225–26.

49. *Comptes rendus de l'Association Géodésique Internationale* (1901), 25 September—6 October 1900, session of 4 October; also Bassot, "Revision de l'arc" (1900), 1275.

50. *Comptes rendus de l'Académie des Sciences* 134 (1902); 965–66, 968, 969, and 970.

51. *Comptes rendus de l'Académie des Sciences* 136 (1903), 861.

52. *Comptes rendus de l'Académie des Sciences* 136 (1903), 861–62.

53. *Comptes rendus de l'Académie des Sciences* 136 (1903), 862 and 868; on the destruction, *Comptes rendus de l'Académie des Sciences* 138 (1904), 1014–15 (Monday 25 April 1904); native informants, *Comptes rendus de l'Académie des Sciences* 140 (1905), 998 and 999 (Monday 10 April 1905); quotation from *Comptes rendus de l'Académie des Sciences* 136 (1903), 871.

54. Laurent Rollet, *Henri Poincaré. Des Mathématiques à la Philosophie. Étude du parcours intellectuel, social et politique d'un mathématicien au début du siècle,* unpublished doctoral dissertation, University of Nancy 2, 1999, 165.

55. Poincaré, "Sur les Principes de la mécanique"; originally from *Bibliothèque du Congrès international de philosophie* III, Paris (1901), 457–94; modified and reprinted in Poincaré, *Science and Hypothesis* [1902], ch. 6, 90.

56. Poincaré, "The Classic Mechanics," in *Science and Hypothesis* [1902], ch. 6, 110, 104–05.

57. Poincaré, "Hypotheses in Physics"; originally "Les relations entre la physique expérimentale et la physique mathématique" in *Revue générale des sciences pures et appliquées* 11 (1900), 1163–75; reprinted in *Science and Hypothesis* [1902], ch. 9, 144.

58. Poincaré, "Intuition and Logic in Mathematics"; originally "Du rôle de l'intuition et de la logique en mathématiques," in *Comptes Rendus du deuxième Congrès international des mathématiciens tenu à Paris du 6–12 août 1900*; reprinted in Poincaré, *Foundations of Science* (1982), 210–11.

59. Poincaré, "La théorie de Lorentz" [1900], 464.

60. Lorentz, *Versuch einer Theorie* [1895].

61. Poincaré, *Electricité et optique* [1901], 530–32.

62. Christie to Poincaré, 3 August 1899, accompanied by Christie to Loewy, 1 December 1898, and Christie to Colonel Bassot (Directeur du Service Géographique de l'armée), 9 February 1899. Observatoire de Paris, ref. X5, C6. Poin-

caré to Christie, 23 June 1899, 9 August 1899, and undated (but probably shortly after 9 August 1899), all from Christie Papers, Cambridge University Archives, MSS RGO 7/261.

63. Poincaré, "La théorie de Lorentz" [1900], 483.

64. Suppose B sends a signal to A at noon across the distance from A to B, AB. B should set his clock at noon plus the transmission time (the usual procedure). But the speed in the left-going direction is $c + v$, so the transmission speed in the headwind direction t(headwind) is $AB/(c + v)$, and, conversely, transmission time in the tailwind direction is $AB/(c - v)$. The "true" time of transmission from B to A is half the round-trip time, $1/2[t(\text{headwind}) + t(\text{tailwind})]$, while the apparent time of transmission is just t(tailwind). So the error committed by using the apparent time of transmission is the difference between true and apparent transmission times, or

Error = $1/2[t(\text{headwind}) + t(\text{tailwind})] - t(\text{tailwind})$

Using the definitions of t(tailwind) and t(headwind) above:

Error = $1/2[AB/(c + v) - AB/(c - v)] = 1/2AB(c + v - c + v)/(c^2 - v^2) = \sim ABv/c^2$.

Darrigol rightly points out that most historians of relativity have ignored this clock coordination interpretation of local time, *Electrodynamics* (2000), 359–60; also Miller's wide-ranging *Einstein, Picasso* (2001), 200–15; for further references, see Stachel, *Einstein's Collected Papers*, vol. 2 (1989), 308n.

65. *Poincaré et les Physiciens*, unpubl. correspondence from the Henri Poincaré Archive: Annexe 3, document 205, 31 January 1902.

66. *Poincaré et les Physiciens*, unpubl. correspondence from the Henri Poincaré Archive: Annexe 3, document 205, 31 January 1902.

67. Poincaré, *The Foundations of Science* (1982), 352.

68. See the excellent discussion in Henri Rollet, *Henri Poincaré. Des Mathématiques à la Philosophie. Études du parcours intellectuel, social et politique d'un mathématicien au début du siècle*, unpublished doctoral dissertation, University of Nancy 2, 1999, 249ff, quotation on 263; also Débarbat, "An Unusual Use" (1996).

69. Poincaré, "Le Banquet du 11 Mai" (1903), 63.

70. Ibid., 63–64.

71. Débarbat, "An Unusual Use" (1996), 52.

72. Darboux, Appell, and Poincaré, "Rapport" (1908), 538–49.

73. Poincaré, "The Present State and Future of Mathematical Physics," originally "L'État actuel et l'avenir de la physique mathématique" [24 September 1904, Congress of Arts and Science at St. Louis, Missouri], in *Bulletin des Sciences Mathématiques* 28 (1904), 302–24. Reprinted in Poincaré, *Valeur de la Science* (1904), 123–47, this quotation on 123.

74. Poincaré, "The Present State" [1904], 128.

75. Ibid.

76. Ibid., 133. Translation modified: *"en retard"* should be "offset to a later

time"—the clocks do not run "slow" in the sense of running at a slower rate. Italics added.

77. Ibid., 142 and 146–47. *Poincaré et les Physiciens*, unpubl. correspondence from the Henri Poincaré Archive: document 124, 191–93, letter from Poincaré to Lorentz (n.d.) but sometime shortly after Poincaré's return from Saint-Louis.

78. Lorentz, "Electromagnetic phenomena" (1904).

79. Poincaré, "Les Limites de la loi de Newton" (1953–54), 220 and 222.

80. Poincaré, "La Dynamique de l'électron" [1908], 567.

81. Poincaré, "Sur la Dynamique," reprinted in his *Mécanique Nouvelle* (1906), 22; discussed by Miller, *Einstein* (1981).

82. *Poincaré et les Physiciens*, unpubl. correspondence from the Henri Poincaré Archive: letter 127, Lorentz to Poincaré, 8 March 1906.

CHAPTER 5

1. See the excellent work on time in Switzerland, Messerli, *Gleichmässig, pünktlich, schnell* (1995), esp. ch. 5. For biographical details on Mathias Hipp, see de Mestral, *Pionniers suisses* (1960), 9–34; also Weber and Favre, "Matthäus Hipp" (1897); Kahlert, "Matthäus Hipp" (1989). On Hipp's relation to the astronomer Hirsch, and the myriad ways in which the new technologies of time and simultaneity precision joined the history of experimental psychology to astronomy, see the excellent works by Canales, "Exit the Frog" (2001); Schmidgen, "Time and Noise" (2002); and Charlotte Bigg, *Behind the Lines. Spectroscopic Enterprises in Early Twentieth Century Europe*, unpublished doctoral dissertation, University of Cambridge, 2002.

2. On Hipp, Kahlert, "Matthäus Hipp" (1989). Landes's work, *Revolution in Time* (1983), 237–337, is excellent on the Swiss watch industry, though he focuses on clock production and not on networks.

3. See Favarger, *L'Électricité* (1924), 408–09.

4. "Die Zukunft der oeffentlichen Zeit-Angaben" (12 November 1890); Merle, "Tempo!" (1989), 166–78 cited in Dohrn-van Rossum, *History of the Hour* (1996), 350.

5. Favarger, "Sur la Distribution de l'heure civile" (1902).

6. Ibid., 199.

7. Ibid., 200.

8. Ibid., 201.

9. Kropotkin, *Memoirs* (1989), 287.

10. Favarger, "Sur la Distribution de l'heure civile" (1902), 202.

11. Ibid., 203.

12. Ibid. Newspaper quoted in Jakob Messerli, *Gleichmässig Pünktlich Schnell* (1995), 126.

13. Einstein, "On the Investigation of the State of the Ether" [1895]; Einstein, "Autobiographical Notes" (1969), 53.

14. Urner, "Vom Polytechnikum zur ETH," 19–23.

15. For example, Einstein's notes on Weber's lectures, in *Collected Papers*, vol. 1, 142. Weber's own work ranged over a variety of topics in experimental and applied subjects: temperature dependence of specific heats, the energy distribution law for blackbody radiation, alternating current circuits, and carbon filaments. Editors' note, *Collected Papers* 1, 62; Barkan, *Nernst* (1999), 114–17.

16. Einstein's notes on Weber's lectures, in *Collected Papers (Translation)*, vol. 1, 51–53.

17. Einstein to Mileva Marić, 10 September 1899, item 52, in *Collected Papers (Translation)*, vol. 1, 132–33.

18. Einstein to Mileva Marić, August 1899; Letter 8 in Einstein, *Love Letters* (1992), 10–11; also in *Collected Papers (Translation)*, vol. 1, 130–31. On Einstein's specific knowledge of aspects of electrodynamics, see Holton, *Thematic Origins* (1973); Miller, *Einstein's Relativity* (1981); and Darrigol, *Electrodynamics* (2000).

19. On Einstein and the ether, see editors' contribution, "Einstein on the Electrodynamics of Moving Bodies," in *Collected Papers*, vol. 1, 223–25, and Darrigol, *Electrodynamics* (2000), 373–80. On Einstein's early use of the relativity principle, ibid., 379.

20. Einstein to Mileva Marić, May 1901, item 111, in *Collected Papers (Translation)*, vol. 1, 174.

21. Einstein to Mileva Marić, June 1901, item 112, in *Collected Papers (Translation)*, vol. 1, 174–75.

22. Einstein to Jost Winteler, 8 July 1901, item 115, in *Collected Papers (Translation)*, vol. 1, 176–77. On Einstein's battle, see Renn's excellent article, "Controversy with Drude" (1997), 315–54.

23. Department of Internal Affairs to Einstein, 31 July 1901, item 120, in *Collected Papers (Translation)*, vol. 1, 179.

24. Einstein to Marcel Grossmann, September 1901, item 122, in *Collected Papers (Translation)*, vol. 1, 180–81.

25. Einstein to Swiss Patent Office, 18 December 1901, item 129, in *Collected Papers (Translation)*, vol. 1, 188.

26. Einstein to Mileva Marić, 19 December 1901, item 130, in *Collected Papers (Translation)*, vol. 1, 188–89.

27. Einstein to Mileva Marić, 17 December 1901, item 128, in *Collected Papers (Translation)*, vol. 1, 186–87.

28. Einstein to Mileva Marić, 19 December 1901, item 130, in *Collected Papers (Translation)*, vol. 1, 188–89. Translation modified.

29. Einstein to Mileva Marić, 28 December 1901, item131, in *Collected Papers (Translation)*, vol. 1, 189–90.

30. Einstein to Mileva Marić, 4 April 1901, item 96, in *Collected Papers (Translation)*, vol. 1, 162–63.

31. Einstein, advertisement for private lessons, 5 February 1902, item 135, in *Collected Papers (Translation)*, vol. 1, 192.

32. Einstein to Mileva Marić, February 1902, item 136, in *Collected Papers (Translation)*, vol. 1, 192–93.

33. Solovine, introduction to Einstein, *Letters to Solovine* (1993), 9.

34. See Holton, *Thematic Origins* (1973), ch. 7.

35. Mach, *Science of Mechanics* [1893], 272–73.

36. Einstein, "Ernst Mach," 1 April 1916, document 29, in *Collected Papers*, vol. 6, 280.

37. Pearson, *Grammar of Science* [1892], 204, 226, and 227.

38. Einstein, *Letters to Solovine* (1983), 8–9. Mill, *System of Logic* (1965), 322.

39. Einstein, *Letters to Solovine* (1983), 8–9; On Einstein and Mach, see Holton, *Thematic Origins* (1973), ch. 7. Other works that Solovine recalls the group discussing were those by Ampère, *Essai* (1834); Mill, *System of Logic* (1965); and Pearson, *Grammar of Science* [1892]. Poincaré, *Wissenschaft und Hypothese* (1904).

40. Poincaré, *Wissenschaft und Hypothese* (1904), 286–89.

41. Einstein to Schlick, 14 December 1915, document 165, in *Collected Papers*, vol. 8a, 221; *Collected Papers (Translation)*, 161.

42. Most often the Lindemanns rendered Poincaré's *convention* by *Übereinkommen*; but, for example, when Poincaré argued against the philosopher E. Le Roy in *Science and Hypothesis* (p. xxiii) that changed. Poincaré wrote, "Some people have been struck by this characteristic of free convention [French: "*de libre convention*," *Science et Hypothèse* (24)] which may be recognized in certain fundamental principles of the sciences." The Lindemanns put the relevant phrase as " . . . *den Charakter freier konventioneller Festsetzungen . . .*" (Poincaré, *Wissenschaft und Hypothese* (1904), XIII).

43. Einstein to Solovine, 30 October 1924, in Einstein, *Letters to Solovine*, 63. One thinks (inter alia) of Max Planck, the first famous theorist-supporter of relativity, who disdained "convenience" talk in physics, celebrating instead that which was universal and invariant. See, for example, Heilbron, *Dilemmas* (1996), 48–52.

44. Einstein, "Autobiographische Skizze," in Seelig, ed., *Helle Zeit, Dunkle Zeit* (1956), 12. Einstein to Mileva Marić, February 1902, item 137, in *Collected Papers (Translation)*, vol. 1, 193. On 2 June 1902 Einstein was notified officially that he had the job at the Patent Office at an annual salary of Sfr 3,500: item 140, in *Collected Papers (Translation)*, vol. 1, 194–95; see also item 141, 19 June 1902, 195; and that he was to assume his duties on 1 July 1902: item 142, 196.

45. On Poincaré's views on dynamics and kinematics, see Miller, *Frontiers* (1986), parts I, III; Paty, *Einstein Philosophe* (1993), 264–76; and Darrigol, *Electrodynamics* (2000).

46. Here is not the place to offer a full technical reconstruction of all aspects of Einstein's path to special relativity. The reader is referred to an excellent short synthesis in Stachel et al., "Einstein on the Special Theory of Relativity," editorial note in *Collected Papers*, vol. 2, 253–74, esp. 264–65. For further development of early history, see, for example, Miller, *Einstein's Relativity* (1981); Darrigol, *Electrodynamics* (2000); and Pais, *Subtle Is the Lord* (1982).

47. Flückiger, *Albert Einstein* (1974), 58.

48. Einstein to Hans Wohlwend, 15 August–3 October 1902, item 2, in *Collected Papers (Translation)*, vol. 5, 4–5.

49. Flückiger, *Albert Einstein* (1974), 58.

50. Flückiger, *Albert Einstein* (1974), 67; Einstein to Mileva Marić, September 1903, item 13, in *Collected Papers (Translation)*, vol. 5, 14–15.

51. References cited in Pais, *Subtle Is the Lord* (1982), 47–48, who cites Flückiger, *Albert Einstein* (1974).

52. Nicolas Stoïcheff, patent 30224 submitted 6 January 1904 issued 1904; American Electrical Novelty, "Stromschliessvorrichtung an elektrischen Pendelwerken," patent 31055 submitted 16 March 1904 and issued 1905.

53. According to the lists of patents provided in Berner, *Initiation* (c. 1912), ch. 10.

54. Hundreds of relevant patents are listed in the *Journal Suisse d'horlogerie* during the relevant years (1902–05). Sadly, the Swiss patent office dutifully destroyed all papers processed by Einstein 18 years after their creation; this was standard procedure on patent opinions, and even Einstein's fame led to no exception. See Fölsing, *Einstein* (1997), 104.

55. The most detailed linkage between Einstein's patent work and his scientific work is on gyromagnetic compasses and Einstein's production of the Einstein-de Haas effect, see Galison, *How Experiments End* (1987), ch. 2; in addition, see Hughes, "Einstein" (1993), and Pyenson, *Young Einstein* (1985). On Einstein's assignment to evaluate electrical patents, see Flückiger, *Albert Einstein* (1974), 62.

56. Flückiger, *Albert Einstein* (1974), 66.

57. J. Einstein & Co. und Sebastian Kornprobst, "Vorrichtung zur Umwandlung der ungleichmässigen Zeigerausschläge von Elektrizitäts-Messern in eine gleichmässige, gradlinige Bewegung," Kaiserliches Patentamt 53546, 26 February 1890; idem, "Neuerung an elektrischen Mess- und Anzeigervorrichtungen," Kaiserliches Patentamt 53846, 21 November 1889; idem, "Federndes Reibrad", 60361, 23 February 1890; "Elektrizitätszähler der Firma J. Einstein & Cie., München (System Kornprobst)" (1891), 949. Also see Frenkel and Yavelov, *Einstein* (in Russ-

ian) (1990), 75 ff., and Pyenson, *Young Einstein* (1985), 39–42. Further on the links between electric clocks and electricity measuring devices, see, for example, Max Moeller, "Stromschlussvorrichtung an elektrischen Antriebsvorrichtungen fuer elektrische Uhren, Elektrizitätszähler und dergl." (Swiss patent 24342).

58. Swiss Patent Office to Einstein, 11 December 1907, item 67, in *Collected Papers (Translation)*, vol. 5, 46. On Einstein's interest in dynamos, see Miller, *Frontiers of Physics* (1986), ch. 3. Einstein insisted that he only took up the role of expert witness if he judged the side he was defending to be in the right. For example, in 1928 he defended Siemens & Halske against Standard Telephones & Cables Ltd.—see Hughes, "Inventors" (1993), 34.

59. Galison, *How Experiments End* (1987), ch. 2.

60. Einstein to Heinrich Zangger, 29 July 1917, document 365, in *Collected Papers*, vol. 8a, 495–96.

61. Paul Habicht to Einstein, 19 February 1908, item 86, in *Collected Papers (Translation)*, vol. 5, 58–61, on 60.

62. On "the little machine" see editors' essay in *Collected Papers*, vol. 5, 51–54; Fölsing, *Einstein* (1997), 132, 241, and 267–78; Frenkel and Yavelov, *Einstein* (1990), ch. 4.

63. Einstein to Albert Gockel, March 1909, item 144, in *Collected Papers (Translation)*, vol. 5, 102.

64. Einstein to Conrad Habicht, 24 December 1907, item 69 in *Collected Papers (Translation)*, vol. 5, 47; Einstein to Jakob Laub, after 1 November 1908, item 125 in *Collected Papers (Translation)*, vol. 5, 90. "I am presently carrying on an extremely interesting correspondence with H. A. Lorentz on the radiation problem. I admire this man like no other; I might say, I love him." Einstein to Jakob Laub, 19 May 1909, item 161 in *Collected Papers (Translation)*, vol. 5, 120–22, on 121.

65. C. Vigreux and L. Brillié, "Pendule avec dispositif électro-magnétique pour le réglage de sa marche," patent 33815.

66. Einstein, "How I Created the Theory" (1982). Compare remarks that Einstein recorded on a discograph in 1924: "After seven years of reflection in vain (1898–1905), the solution came to me suddenly with the thought that our concepts and laws of space and time can only claim validity insofar as they stand in a clear relation to our experiences; and that experience could very well lead to the alteration of these concepts and laws. By a revision of the concept of simultaneity into a more malleable form, I thus arrived at the special theory of relativity." Einstein, in *Collected Papers (Translation)*, vol. 2, 264.

67. Joseph Sauter, "Comment j'ai appris à connaître Einstein," printed in Flückiger, *Albert Einstein in Bern* (1972), 156; Fölsing, *Albert Einstein* (1997), 155–56.

68. Einstein to Habicht, May 1905, document 27, in *Collected Papers (Translation)*, vol. 5, 19–20, on 20.

69. Einstein, "Conservation of Motion" [1906] in *Collected Papers (Translation)*, vol. 2, 200–206, on 200.

70. Cohn, "Elektrodynamik" (1904).

71. Einstein, "Relativity Principle" [1907], document 47, in *Collected Papers*, vol. 2, 432–88, on 435; *Collected Papers (Translation)*, 252–311, on 254 (translation modified); also Einstein to Stark, 25 September 1907, document 59 in *Collected Papers*, vol. 5, 74–75. On Cohn's physics, cf. Darrigol, *Electrodynamics* (2000), 368, 382, and 386–92.

72. Abraham, *Theorie der Elektrizität: Elektromagnetische Theorie der Strahlung* (Leipzig, 1905), 366–79; cited in Darrigol, *Electrodynamics* (2000), 382.

73. Warwick, "Cambridge Mathematics" (1992, 1993).

74. Cited in Galison, "Minkowski" (1979), 98, 112–13.

75. Galison, "Minkowski" (1979), 97.

76. For an excellent overview of the reception of Minkowski's ideas, see Walter, "The non-Euclidean style" (1999).

77. Cited in Galison, "Minkowski" (1979), 95.

78. Einstein, "The Principle of Relativity and Its Consequences in Modern Physics" [1910], document 2, in *Collected Papers (Translation)*, vol. 3, 117–43, on 125.

79. Einstein, "The Theory of Relativity" [1911], document 17, in *Collected Papers (Translation)*, vol. 3, 340–50, on 348 and 350.

80. "Discussion" Following Lecture Version of "The Theory of Relativity," document 18, in *Collected Papers (Translation)*, vol. 3, 351–58, on 351–52.

81. "Discussion" Following Lecture Version of "The Theory of Relativity," document 18, in *Collected Papers (Translation)*, vol. 3, 351–58, on 356–58; see also the notes to the original version in *Collected Papers*, vol. 3, on 448–49. Poincaré, *La Science* (1905), on 165.

82. Laue, *Relativitätsprinzip* (1913), 34.

83. Planck, *Eight Lectures* (1998), 120; translation modified slightly following Walter, "Minkowski" (1999), 106. On Planck's words and Einstein's career, see Illy, "Albert Einstein," 76.

84. Einstein, "On the Principle of Relativity" [1914], document 1, in *Collected Papers*, vol. 6, 3–5, on 4; *Collected Papers (Translation)*, vol. 6, 4.

85. Cohn, "Physikalisches" (1913), 10; Einstein, "On the Principle of Relativity" [1914], document 1, in *Collected Papers (Translation)*, vol. 6, 4.

86. More precisely, this simple triangle shows quantitatively the relation between time measured in the two frames. Let Δt be the time it takes light to travel the distance h (so $h = c\Delta t$). We imagine the light clock moving to the right at a

speed v, and the inclined path of the light pulse to cover a distance D in a time $\Delta t'$ (so $D = c\Delta t'$). In the time $\Delta t'$ that the beam takes to reach the top mirror, the launching point of the light beam has moved to the right an amount b which must be the speed of the clock multiplied by the time $\Delta t'$, that is: $b = v\Delta t'$. So we have a right triangle (see 5.12b) with all the sides given. Apply Pythagorus's theorem: $D^2 = b^2 + h^2$. Substitute the values of D, b, and h, and we have: $(c\Delta t')^2 = (v\Delta t')^2 + (c\Delta t)^2$. Subtracting $(v\Delta t')^2$ from both sides and simplifying, we get: $\Delta t' / \Delta t = 1/\sqrt{(1-v^2/c^2)}$. This is *the* crucial result. It says that a tick of the clock at rest that takes Δt in its rest frame (here light moving a distance h) is measured by an observer at rest to take a longer time ($\Delta t'$) when the clock is moving at velocity v. For $v/c = 4/5$ the speed of light, this ratio $1/\sqrt{(1-v^2/c^2)}$ is 5/3: a clock moving at four-fifths the speed of light would be measured by a stationary clock to run slow by a factor of 5/3. Of course "stationary" and "in motion" are, according to Einstein, entirely relative.

87. Howeth, *History* (1963). Further on wireless time setting, see, for example, Roussel, *Premier Livre* (1922), esp. 150–52. Boulanger and Ferrié, *La Télégraphie* (1909), date the Eiffel Tower radio station to 1903. Ferrié, "Sur quelques nouvelles applications de la Télégraphie" (1911), esp. 178, indicates that planning for wireless time coordination began at the start of work on wireless; Rothé, *Les Applications de la Télégraphie* (1913), discusses the details of radio-communicated time coordination procedure.

88. Max Reithoffer and Franz Morawetz, "Einrichtung zur Fernbetätigung von elektrischen Uhren mittels elektrischer Wellen," Swiss patent 37912, submitted 20 August 1906.

89. Depelley, *Les Cables sous-marins* (1896), 20.

90. Poincaré, "Notice sur la télégraphie sans fil" [1902].

91. Amoudry, *Le Général Ferrié* (1993), 83–95.

92. Conférence Internationale de l'heure, in *Annales du Bureau des Longitudes* 9, D17.

93. Commission Technique Interministérielle de Télégraphe sans Fil, 7th Meeting, 8 March 1909, MS 1060, II F1, Archives of the Paris Observatory.

94. Poincaré "La Mécanique nouvelle" (1910), 4, 51, 53–54.

95. (Approval) Ministre de l'Instruction publique et des Beaux-Arts, à Monsieur le Directeur de l'Observatoire de Paris, 17 July 1909; (Meeting minutes) Commission Technique Interministérielle de Télégraphe sans Fil, 10th Meeting, 26 June 1909, both documents MS 1060, II F1, Archives of the Paris Observatory.

96. Poincaré, "La Mécanique Nouvelle" [1909], 9. Manuscript copy is dated 24 July 1909 (Archives de l'Académie des Sciences).

97. Poincaré, "La Mécanique Nouvelle," Tuesday 3 August 1909.

98. Amoudry, *Général Ferrié* (1993), 109; see Comptes rendus de l'Académie des Sciences report 31 January [1910].

99. *Scientific American* 109, 13 December 1913, 455; see also Joan Marie Mathys (unpubl. MA thesis, "The Right Place at the Right Time," Marquette University, 30 September 1991).

100. Lallemand, "Projet d'organisation d'un service international de l'heure" (1912). On the Eiffel Tower-Arlington exchange, see, for example, Amoudry, *Général Ferrié* (1993), 117; Joan Marie Mathys (unpublished MA thesis, George Washington University, 1991); and *Scientific American* 109, 13 December 1913, 445.

101. Howse, *Greenwich* (1980), 155.

102. See 11th and 13th meetings of the Commission Technique Interministérielle de Télégraphe sans Fil, 21 March 1911 and 21 November 1911, MS 1060, II F1, Archives of the Paris Observatory.

103. Léon Bloch, *Le Principe de la relativité* (1922), 15–16. Dominique Pestre characterizes Bloch (and his brother) as physicists who were unusual for their time in France by writing textbooks that looked positively on the new physics of the early twentieth century and characteristically wrote using a series of progressive generalizations from the concrete to the abstract (no doubt to appeal to their more experimentally oriented colleagues). See Pestre, *Physique et Physiciens* (1984), 18, 56, and 117.

104. See Bureau des Longitudes, *Réception des signaux horaires: Renseignements météorologiques, seismologiques, etc., transmis par les postes de télégraphie sans fil de la Tour Eiffel*, Lyon, Bordeaux, etc. (Bureau des Longitudes, Paris, 1924), 83–84.

105. Corrections are of many sorts, and include effects due to the motion of satellites, the lower gravitational field at the height of the satellites, and the rotatory motion of the earth. The relativistic component of the Doppler shift is $v^2/2c^2$, which for satellite velocities is about seven millionths of a second per day. Because the speed of light is so much faster than the speed of the satellites, most of general relativity need not be taken into account, but the equivalence principle that is part of general relativity is significant. (The equivalence principle says that there is no way to distinguish the physics of a freely falling box from the same box without a gravitational field.) A more rigorous analysis would take into account (inter alia) that the satellites' orbit is not always in the same gravitational field, that the earth observer may be moving on the surface of the earth, that the earth's gravitational field is not the same everywhere on the earth's surface, that the sun's gravitational field affects the earth clock and the satellite clock differently, and that the apparent velocity of light is altered by the earth's gravitational field.

106. Neil Ashby, "General Relativity in the Global Positioning System," www.phys.lsu.edu/mog/mog9/node9.html (accessed 28 June 2002).

107. Chronology of activist actions, www.plowshares.se/aktioner/plowcron5.htm (accessed 19 February 2002); also see *Los Angeles Times*, 12 May 1992, "Men Arrested in Space Satellite Hacking Called Peace Activists," Metro part B, 12.

108. Taylor, "Propaganda by Deed" (n.d.), 5, in "Greenwich Park Bomb file," Cambridge University archives. Serge F. Kovaleski, "1907 Conrad Novel May Have Inspired Unabomb Suspect," *Washington Post*, 9 July 1996, A1.

109. This has been pointed out many times: Infeld, *Albert Einstein* (1950), 23; Holton, *Thematic Origins* (1973); and Miller, *Einstein's Relativity* (1981), and idem, "The Special Relativity Theory" (1982), 3–26. Einstein does refer to "Lorentz's theory" in the text (but no footnote).

110. Myers, "From Discovery to Invention" (1995), 77.

111. By law, patent officers were taught to look for originality. In Switzerland that hunt for novelty had a specific meaning: "Discoveries do not count as new, if at the time of their registration in Switzerland, they are well enough known that their development by the technically adept is already possible." The contrast with neighboring countries is worth noting. In France rejection for lack of originality is based on "publicity" given to prior work, while in Germany a similar refusal could be made if the invention had been either reported in an official publication of the last century *or* if its use was so well known that its employment by other technical people appeared possible. According to turn-of-the-century patent manuals, Swiss law lay closer to the French; originality in Switzerland meant that the invention was not actually known in Switzerland regardless of what might lie dormant in an obscure foreign publication.

112. Schanze, *Patentrecht* (1903), 33.

113. Ibid., 33–34.

114. Einstein, "The World as I See It," in idem, *Ideas and Opinions* (1954), 10.

Chapter 6

1. Einstein, "On the Relativity Principle and the Conclusions Drawn From It" [1907], document 47, in *Collected Papers*, vol. 2, 432–88; in *Collected Papers (Translation)*, vol. 2, 252–55.

2. Einstein did cite Poincaré's work on the inertia of energy in Einstein's 1906 derivation of $E = mc^2$ ("The Principle of Conservation of Motion of the Center of Gravity" [1906], document 35, in *Collected Papers (Translation)*, vol. 2, 200–206); then in ("On the Inertia of energy" [1907], document 45, in *Collected Papers (Translation)*, vol. 2, 238–50). Einstein re-omitted Poincaré when Einstein again wrote on the inertia of energy.

3. Poincaré, "La Mécanique Nouvelle" [1909], 9. Manuscript copy is dated 24 July 1909 (Archives de l'Académie des Sciences).

4. Faguet, *Après l'Ecole* (1927), 41.

5. Einstein, "On the Development of our Views Concerning the Nature and Constitution of Radiation," document 60, in *Collected Papers (Translation)*, vol. 2, 379.

6. "Discours du Duc M. de Broglie" in Poincaré, *Livre du centenaire* (1935), 71–78, on 76.

7. General conclusions by Poincaré, in Langevin and de Broglie, *Théorie du rayonnement* (1912), 451.

8. I have retranslated this quotation and also corrected a spurious insertion that seems to have crept into the secondary literature. The phrase *gegen die Relativitaetstheorie* simply does not appear in the original. Einstein to Heinrich Zangger, 15 November 1911, item 305, in *Collected Papers*, vol. 5, 249–50.

9. Poincaré to Weiss, editors of Poincaré Papers date as ca. November 1911. *Poincaré et les Physiciens*, unpubl. correspondence from the Henri Poincaré Archive Zürich.

10. Darboux, "Eloge historique" [1913], lxvii.

11. Poincaré, "Space and Time" (1963), 18 and 23.

12. Ibid., 24.

13. Poincaré, "Moral Alliance," *Last Essays* (1963), 114–17, on 114, 117; on the last days of Poincaré, Darboux, "Éloge" (1916). See the excellent discussion of Poincaré's political engagement in Laurent Rollet, *Henri Poincaré. Des Mathématiques à la Philosophie. Étude du parcours intellectuel, social et politique d'un mathématicien au début du siècle*, unpublished doctoral dissertation, University of Nancy 2, 1999, esp. 283–84.

14. "Discours du Prince Louis de Broglie" (1955), 66.

15. Poincaré, *Foundations of Science* (1982), 352 (translation slightly modified).

16. Ibid., 232.

17. Sherman, *Telling Time* (1996).

18. "Lettre de M. Pierre Boutroux à M. Mittag-Leffler" [18 June 1913], in Poincaré, *Oeuvres*, vol. 11 (1956), 150.

19. Both quotations, cf. Einstein, *Ideas and Opinions* (1954), 274.

20. Einstein about the religious spirit of science in *Mein Weltbild* [1934], reprinted in idem, *Ideas and Opinions* (1954), 40.

21. Editors' introduction to vol. 2, Einstein, *Collected Papers*, xxv–xxvi.

22. Einstein to Schlick, 21 May 1917, *Collected Papers (Translation)*, vol. 8, 333. Also as Holton has argued in *Thematic Origins* (1973) that Einstein's emphasis on metaphysics changed over time.

23. Favarger, *L'Électricité* (1924), 10.

24. Ibid., 11.

25. Representations of Einstein's relativity as a culmination of increasingly accurate "no-aether" measurements are rife; perhaps the most scholarly attempt to locate Einstein's formulation as a mere variant of the early aether-electron theories is to be found in Edmund Whittaker, *History of the Theories of Aether* (1987), 40, where the chapter "The Relativity Theory of Poincaré and Lorentz" includes the remark: "Einstein published a paper [in 1905] which set forth the relativity theory

of Poincaré and Lorentz with some amplifications, and which attracted much attention. He asserted as a fundamental principle the constancy of the speed of light . . . which at the time was widely accepted, but has been severely criticized by later writers." See Holton, *Thematic Origins* (1973), esp. ch. 5 and Miller, *Einstein's Relativity* (1981).

26. Schaffer, "Late Victorian metrology" (1992), 23–49; Wise, "Mediating Machines" (1988); Galison, *Image and Logic* (1997).

27. Einstein to Solovine, Princeton, 3 April 1953, in Einstein, *Letters to Solovine* (1987), 143; translation corrected.

28. Einstein to the son and sister of Michele Besso, 21 March 1955, *Albert Einstein Michele Besso Correspondance* (1972), 537–39.

BIBLIOGRAPHY

Abraham, M. 1905. *Theorie der Elektrizität: Elektromagnetische Theorie der Strahlung*, Vol. II. Leipzig: B. G. Teubner.

Alder, Ken. 2002. *The Measure of All Things*. New York: The Free Press.

Allen, John S. 1951. *Standard Time in America. Why and How it Came about and the Part Taken by the Railroads and William Frederick Allen*. New York: National Railway Publication Company.

Allen, William F. 1883. *Report on the Subject of National Standard Time Made to the General and Southern Railway. Time Conventions*. New York: National Railway Publication Company.

Allen, William F. 1884. *History of the Adoption of Standard Time Read before the American Metrological Society on December 27th 1883, With other Papers Relating Thereto*. New York: American Metrological Society.

Ambronn, Friedrich A. L. 1889. *Handbuch der astronomischen Instrumentenkunde. Eine Beschreibung der bei astronomischen Beobachtungen benutzten Instrumente sowie Erläuterung der ihrem Bau, ihrer Anwendung und Aufstellung zu Grunde liegenden Principien*, Vol. I. Berlin: Verlag von Julius Springer.

Amoudry, Michel. 1993. *Le général Ferrié. La naissance des transmissions et de la radiodiffusion*. Préface de Marcel Bleustein-Blanchet. Grenoble: Presses universitaires de Grenoble.

Ampère, André-Marie. 1834. *Essai sur la philosophie des sciences. Ou exposition analytique d'une classification naturelle de toutes les connaissances humaines*. Paris: Bachelier.

Andersson, K. G. 1994. "Poincaré's Discovery of Homoclinic Points." In *Archive for History of Exact Sciences*, Vol. 48, pp. 133–47.

Andrade, Jules. 1898. *Leçons de mécanique physique*. Paris: Société d'éditions scientifiques.

Aubin, David. "The Fading Star of the Paris Observatory in the Nineteenth Century: Astronomers' Urban Culture of Circulation and Observation." In *Science and the City* (Osiris 18), eds. Sven Dierig, Jens Lachmund, and Andrew Mendelsohn (forthcoming).

Baczko, Bronislaw. 1992. "Le Calendrier républicain." In *Les lieux de mémoire* [1984], Vol. 1, ed. Pierre Nora. Paris: Gallimard.

Barkan, Diana Kormos. 1993. "The First Solvay. The Witches' Sabbath: The First International Solvay Congress of Physics." In *Einstein in Context* 6, pp. 59–82.

Barkan, Diana Kormos. 1999. *Walther Nernst and the Transition to Modern Science*. Cambridge: Cambridge University Press.

Barnard, F.A.P. 1882. "A Uniform System of Time Reckoning." In *The Association for the Reform and Codification of the Law of Nations. The Committee on Standard Time. Views of the American Members of the Committee, As to the Resolutions Proposed at Cologne Recommending a Uniform System of Time Regulation for the World*. New York: Moggowan & Slipper, pp. 3–4.

Barnard, F.A.P., 1884. "The Metrology of the Great Pyramid." In *Proceedings of the American Metrological Society* 4, pp. 117–219.

Barrow-Green, June. 1997. "Poincaré and the Three Body Problem." In *History of Mathematics* 11. Providence, RI: American Mathematical Society.

Barthes, Roland. 1972. *Mythologies*. Selected and Translated from the French by Annette Lavers. London: Paladin Grafton Books, reprinted 1989.

Bartky, Ian R. 1983. "The Invention of Railroad Time." In *Railroad History Bulletin* 148, pp. 13–22.

Bartky, Ian R. 1989. "The Adoption of Standard Time." In *Technology and Culture* 30 (1), pp. 25–57.

Bartky, Ian R. 2000. *Selling the True Time. Nineteenth-Century Timekeeping in America*. Stanford: Stanford University Press.

Bennett, Jim. 2002. "The Travels and Trials of Mr. Harrison's Timekeeper." In *Instruments, Travel, and Science*, eds. Marie-Noëlle Bourguet, Christian Licoppe, and H. Otto Sibum. New York: Routledge.

Bergson, Henri. 2001. *Time and Free Will*. New York: Dover.

Bernardières, O. de. 1884. "Déterminations télégraphiques de différences de longitude dans l'Amérique du Sud." In *Comptes Rendus de l'Academie des Sciences* 98, pp. 882–90.

Berner, Albert. ca. 1912. *Initiation de l'horloger à l'électricité et à ses applications*. Préface de L. Reverchon. La Chaux-de-Fonds, Switzerland: "Inventions-Revue."

Bernstein, A. 1897. *Naturwissenschaftliche Volksbücher*. Berlin: Ferd. Dummlers Verlagsbuchhandlung.

Blaise, Clark. 2000. *Time Lord. Sir Sandford Fleming and the Creation of Standard Time*. London: Weidenfeld & Nicolson.

Bloch, Léon. 1922. *Principe de la relativité et la théorie d'Einstein*. Paris: Gauthier-Villars et Fils.

Bond, W.C. 1856. "Report to the Director" [4 December 1850]. In *Annals of the Astronomical Observatory of Harvard College* 1, pp. cxl–cl.

Boulanger, Julien A., and G. A. Ferrié. 1909. *La Télégraphie sans fil et les ondes électriques*. Paris: Berger-Levrault et Fils.

Breguet, Antoine. 1880–1881. "L'unification de l'heure dans les grandes villes." In *Le Génie civil. Revue générale des industries Françaises et étrangères* 1, pp. 9–11.

Broglie, Louis-Victor de. 1955. "Discours du Prince de Broglie." In *Le livre du centenaire*. Paris: Gauthier-Villars et Fils, pp. 62–71.

Broglie, Maurice de. 1955. "Discours du Duc Maurice de Broglie. Henri Poincaré et la Philosohie." In *Le Livre du centenaire*. Paris: Gauthier-Villars et Fils, pp. 71–78.

Bucholz, Arden. 1991. *Moltke, Schlieffen, and Prussian War Planning*. New York, Oxford: Berg.

Bucholz, Arden. 2001. *Moltke and the German Wars, 1864–1871*. New York: Palgrave.

Burnett, D. Graham. 2000. *Masters of All They Surveyed*. Chicago: The University of Chicago Press.

Burpee, Lawrence J., and Sandford Fleming. 1915. *Empire Builder*. Oxford: Oxford University Press.

Cahan, David. 1989. *An Institute for an Empire. The Physikalisch-Technische Reichsanstalt 1871–1918*. Cambridge: Cambridge University Press.

Calaprice, Alice (ed.). 1996. *The Quotable Einstein*. (With a Foreword by Freeman Dyson.) Princeton, NJ: Princeton University Press.

Calinon, A. 1885. "Étude critique sur la mécanique." In *Bulletin de la Société de Sciences de Nancy* 7, pp. 76–180.

Calinon, A. 1897. *Étude sur les diverses grandeurs en mathématiques*. Paris: Gauthier- Villars et Fils.

Canales, Jimena. 2001. "Exit the Frog: Physiology and Experimental Psychology in Nineteenth-Century Astronomy." In *British Journal for the History of Science* 34, pp. 173–97.

Canales, Jimena. 2002. "Photogenic Venus. The 'Cinematographic Turn' and its Alternatives in Nineteenth-Century France." In *Isis* 93, pp. 585–613.

Cantor, G.N., and M.J.S. Hodge (eds.). 1981. *Conceptions of Ether*. Cambridge: Cambridge University Press.

Cassidy, David. 2001. *Uncertainty. The Life and Science of Werner Heisenberg*. New York: Freeman.

Cassidy, David. 2001. "Understanding the History of Special Relativity." In *Science and Society: The History of Modern Physical Science in the Twentieth Century,*

Vol. 1: Making Special Relativity, eds. P. Galison, M. Gordon, and D. Kaiser. New York: Routledge, pp. 229–47.

Chapuis, Alfred. 1920. *Histoire de la Pendulerie Neuchâteloise (Horlogerie de Gros et de Moyen Volume)*. Avec la collaboration de Léon Montandon, Marius Fallet, et Alfred Buhler. (Préface de Paul Robert.) Paris, Neuchâtel: Attinger Frères.

Christie, K.C.B. 1906. *Telegraphic Determinations of Longitude (Royal Observatory Greenwich). Made in the Years 1888 to 1902*. Edinburgh: Neill & Co.

Cohen, I. Bernard, and Anne Whitman. 1999. *Isaac Newton. The Principia*. California: University of California Press.

Cohn, Emil. 1904. "Zur Elektrodynamik bewegter Systeme." In *Sitzung der physikalisch-mathematischen Classe der Akademie der Wissenschaften*, Vol. 10. pp. 1294–1303, 1404–16.

Cohn, Emil. 1913. *Physikalisches über Raum und Zeit*. Leipzig, Berlin: B.G. Teubner.

Comptes rendus des séances de la Douzième conférence générale de l'association géodésique internationale. Réunie à Stuttgart du 3 au 12 Octobre 1898. Neuchâtel: Paul Attinger, 1899.

Comptes rendus de l'Association Géodésique Internationale, 25 Septembre–6 Octobre 1900 [4 October 1900]. Neuchâtel: Paul Attinger, 1901.

Conférence générale des poids et mesures. *Rapport sur la construction, les comparaisons et les autres opérations ayant servi à déterminer les équations des nouveaux prototypes métriques*. Présenté par le Comité International des Poids et Mesures. Paris: Gauthier-Villars et Fils, 1889.

"Conférence Internationale de l'heure." In *Annales du Bureau des Longitudes* 9, p. D17.

Conrad, Joseph. 1953. *The Secret Agent. A Simple Tale*. Stuttgart: Tauchnitz.

Cornu, M. A. 1894. "La synchronisation électromagnétique (Conférence faite devant la Société internationale des Électriciens le 24 janvier 1894, Paris)." In *Bulletin de la Société internationale des Électriciens* 11, pp. 157–220.

Cornu, M. A. 1897. "La Décimalisation de l'heure et de la circonférence," In *L'Eclairage Electrique* 11, pp. 385–90.

Creet, Mario. 1990. "Sandford Fleming and Universal Time." In *Scientia Canadensis* 14, pp. 66–90.

Darboux, Gaston. 1916. "Éloge historique d'Henri Poincaré." In Poincaré, *Oeuvres*, Vol. 2, pp. VII–LXXI.

Darrigol, Oliver. 2000. *Electrodynamics from Ampère to Einstein*. Oxford: Oxford University Press.

Daston, L. J. 1986. "The Physicalist Tradition in Early Nineteenth Century French Geometry." In *Studies in History and Philosophy of Science* 17, pp. 269–95.

Davis, C. H., J. A. Norris, C. Laird. 1885. *Telegraphic Determination of Longitudes in Mexico and Central America and on the West Coast of South America*. Washington: U. S. Government Printing Office.

Débarbat, Suzanne. 1996. "An Unusual Use of an Astronomical Instrument: The Dreyfus Affair and the Paris 'Macro-Micromètre'." In *Journal for the History of Astronomy* 27, pp. 45–52.

Depelley, M. J. 1896. *Les cables sous-marins et la défense de nos colonies*. Paris: Léon Chailley.

Diacu, Florin, and Philip Homes. 1996. *Celestial Encounters. The Origins of Chaos and Stability*. Princeton, NJ: Princeton University Press.

Documents de la Conférence Télégraphique Internationale de Paris (May–June 1890). *Bureau International des administrations Télégraphiques*. Berne: Imprimerie Rieder & Simmer, 1891.

Documents diplomatiques de la conférence du mètre. Paris: Imprimérie nationale, 1875.

Dohrn-van Rossum, Gerhard. 1996. *History of the Hour. Clocks and Modern Temporal Orders*. Trans. by Thomas Dunlap. Chicago: University of Chicago Press.

Dowd, Charles F. 1930. *A Narrative of His Services in Originating and Promoting the System of Standard Time Which Has Been Used in the United States of America and in Canada since 1883*. Ed. Charles N. Dowd. New York: Knickerbocker Press.

Einstein, Albert. [1895]. "On the Investigation of the State of the Ether in a Magnetic Field." In idem, *Collected Papers*, Vol. 1 (1987), pp. 6–9.

Einstein, Albert. 1905. "Elektrodynamik bewegter Körper." In *Annalen der Physik* 17, pp. 891–921.

Einstein, Albert. [1949]. "Autobiographical Notes." In *Albert Einstein*, Vol. 1 ed. P. A. Schilipp. La Salle, IL: Open Court, 1969, 1992.

Einstein, Albert. 1954. *Ideas and Opinions*. Ed. Carl Seelig. (New translations and revisions by Sonja Bargmann.) New York: Bonanza Books.

Einstein, Albert. 1960. *Einstein on Peace*. Eds. Otto Nathan and Heinz Norden. (Preface by Bertrand Russel.) New York: Avenel Books.

Einstein, Albert. 1982. "How I Created the Theory of Relativity." In *Physics Today* 35, pp. 45–47. Retranslated by Ryoichi Itagaki for the Einstein, *Collected Papers* (forthcoming).

Einstein, Albert. 1987–. *The Collected Papers of Albert Einstein*. Ed. John Stachel et al. Translation by Anna Beck. Princeton: Princeton University Press.

Einstein, Albert. 1993. *Letters to Solovine*. (With an Introduction by Maurice Solovine, 1987.) New York: Carol Publishing Group.

Einstein, Albert, and Michele Besso. 1972. *Correspondence 1903–1955*. Traduction, notes et introduction de P. Speziali. Paris: Hermann.

Ekeland, Ivar. 1988. *Mathematics and the Unexpected*. (With a Foreword by Felix E. Browder.) Chicago: The University of Chicago Press.

Faguet, E., P. Painlevé, E. Perrier, and H. Poincaré. 1927. *Après l'École. Ce que disent les livres. Ce que disent les choses*. Paris: Hachette.

Favarger, M. A. 1884. "L'électricité et ses applications à la chronométrie." In *Journal Suisse d'Horlogerie. Revue Horlogère Universelle* (6), pp. 153–58.

Favarger, M. A. 1902. "Sur la distribution de l'heure civile." In *Comptes rendus des travaux, procès-verbaux, rapports et mémoires*. Paris: Gauthier-Villars et Fils, pp. 198–203.

Favarger, M. A. 1920. "Les horloges électriques." In *Histoire de la Pendulerie Neuchâteloise*, ed. Alfred Chapuis. Paris: Attinger, pp. 399–420.

Favarger, M. A. 1924. *L'Électricité et ses applications à la chronométrie*. Neuchâtel: Édition du journal suisse d'horlogerie et de bijouterie.

Ferrié, A. G. 1911. "Sur quelques nouvelles applications de la télégraphie sans fil." In *Journal de Physique* 5, pp. 178–89.

Fleming, Sandford. 1876. *Terrestrial Time. A Memoir*. London: Edwin S. Boot. (Reprinted by the Canadian Institute for Historical Microreproductions, 1980).

Fleming, Sandford. 1879. "Longitude and Time-Reckoning." In *Papers on Time-Reckoning. From the Proceedings of the Canadian Institute, Toronto*, Vol. 1 (4), pp. 52–63.

Fleming, Sandford. 1891. "General von Moltke on Time Reform." In *Documents Relating to the Fixing of a Standard of Time and the Legalization Thereof*. Printed by Order of Canadian Parliament, Session 1891 (8). Ottawa: Brown Chamberlin, pp. 25–27.

Flückiger, Max. 1974. *Albert Einstein in Bern. Das Ringen um ein neues Weltbild. Eine dokumentarische Darstellung über den Aufstieg eines Genies*. Bern: Paul Haupt.

Fölsing, Albrecht. 1997. *Albert Einstein. A Biography*. Trans. Ewald Osers, 1993. New York: Penguin Books.

French, A. P. 1968. *Special Relativity*. Cambridge, MA: MIT Press.

Frenkel, W. J., and B. E. Yavelov. 1990. *Einstein: Invention and Experiment* (Russian). Moscow: Nauka.

Furet, François, and Mona Ozouf. 1992. *Dictionnaire critique de la Révolution Française. Institutions et créations*. Paris: Flammarion.

Galison, Peter. 1979. "Minkowski's Space-Time: From Visual Thinking to the Absolute World." In *Historical Studies in the Physical Sciences* 10, pp. 85–121.

Galison, Peter. 1987. *How Experiments End*. Chicago: The University of Chicago Press.

Galison, Peter. 1997. *Image and Logic. A Material Culture of Microphysics.* Chicago: The University of Chicago Press.

Galison, Peter, Michael Gordin, and David Kaiser, eds. 2001. *Science and Society: The History of Modern Physical Science in the Twentieth Century, Volume 1: Making Special Relativity.* New York: Routledge.

Giedymin, Jerzy. 1982. *Science and Convention. Essays on Henri Poincaré's Philosophy of Science and the Conventionalist Tradition.* Oxford: Pergamon Press.

Gilain, Christian. 1991. "La théorie qualitative de Poincaré et le problème de l'intégration des équations différentielles." In *La France Mathématique*, ed. H. Gispert (Cahiers d'histoire et de philosophie de sciences 34). Paris: Centre de documentation sciences humaines, pp. 215–42.

Gray, Jeremy. 1997. "Poincaré, Topological Dynamics, and the Stability of the Solar System." In *The Investigation of Difficult Things*, eds. P. M. Harman and Alan E. Shapiro. Cambridge: Cambridge University Press, pp. 503–24.

Gray, Jeremy. 1997. "Poincaré in the Archives—Two Examples." In *Philosophia Scientiae* 2, pp. 27–39.

Gray, Jeremy (ed.). 1999. *The Symbolic Universe.* Oxford: Oxford University Press.

Green, Francis M. 1877. *Report on the Telegraphic Determination of Differences of Longitude in the West Indies and Central America.* Washington: U.S. Government Printing Office.

Green, F. M., C. H. Davis, and J. A. Norris. 1880. *Telegraphic Determination of Longitudes of the East Coast of South America.* Washington: U.S. Government Printing Office.

Grünbaum, Adolf. 1968. "Carnap's Views on the Foundations of Geometry." In Arthur P. Schilpp, *The Philosophy of Rudolf Carnap*, pp. 599–684.

Grünbaum, Adolf. 1968. *Geometry and Chronometry.* In *Philosophical Perspective.* Minneapolis: University of Minnesota Press.

Guillaume, C. E. 1890. "Travaux du Bureau International des Poids et Mesures." In *La Nature*, Ser. 1, pp. 19–22.

Harman, P. M. and Alan E. Shapiro. 1992. *The Investigation of Difficult Things. Essays on Newton and the History of the Exact Sciences in Honour of D. T. Whiteside.* Cambridge: Cambridge University Press.

Hayles, Katherine N. (ed.). 1991. *Chaos and Order. Complex Dynamics in Literature and Science.* Chicago: The University of Chicago Press.

Headrick, Daniel R., 1988. *The Tentacles of Progress. Technology Transfer in the Age of Imperialism, 1850–1940.* New York, Oxford: Oxford University Press.

Heilbron, J. L. (1996). *The Dilemmas of an Upright Man. Max Planck and the Fortunes of German Science.* Cambridge, MA: Harvard University Press. (With a new Afterword, 2000.)

Heinzmann, Gerhard. 2001. "The Foundations of Geometry and the Concept of Motion: Helmholtz and Poincaré." In *Science in Context* 14, pp. 457–70.

"Historical Account of the Astronomical Observatory of Harvard College, From October 1855 to October 1876." In *Annals of the Astronomical Observatory of Harvard* 8 (1877), pp. 10–65.

Holton, Gerald. 1973. *Thematic Origins of Scientific Thought. Kepler to Einstein.* Cambridge, MA: Harvard University Press, revised 1988.

Howard, Michael. 1961. *The Franco-Prussian War. The German Invasion of France, 1870–1871.* London, New York: Routledge, 1979.

Howeth, L.S. 1963. *History of Communications-Electronics in the United States Navy.* (With an Introduction by Chester W. Nimitz.) Washington: U.S. Government Printing Office.

Howse, Derek. 1980. *Greenwich Time and the Discovery of the Longitude.* Oxford: Oxford University Press.

Hughes, Thomas. 1993. "Einstein, Inventors, and Invention." In *Science in Context* 6, pp. 25–42.

Illy, J. 1979. "Albert Einstein in Prague." In *Isis* 70, pp. 76–84.

Infeld, Leopold. 1950. *Albert Einstein. His Work and Its Influence on our World.* New York: Charles Scribner's Sons, revised edition.

International Conference Held at Washington. For the Purpose of Fixing a Prime Meridian and a Universal Day. Washington: Gibson Bros., 1884.

Janssen, M. J. 1885. "Sur le congrès de Washington et sur les propositions qui y ont été adoptées touchant le premier Méridien, l'heure universelle et l'extension du système décimal à la mesure des angles et à celle du temps." In *Comptes rendus de l'Académie des Sciences* 100, pp. 706–29.

Jones, Z., and L. G. Boyd. 1971. *The Harvard College Observatory: The First Four Directorships 1839–1919.* Cambridge, MA: Belknap Press.

Kahlert, Helmut. 1989. *Matthäus Hipp in Reutlingen. Entwicklungsjahre eines großen Erfinders (1813–1893). Sonderdruck aus: Zeitschrift für Württembergische Landesgeschichte 48. Hg. von der Kommission für geschichtliche Landeskunde in Baden-Württemberg und dem Württembergischen Geschichts- und Altertumsverein.* Stuttgart: Kohlhammer.

Kern, Stephen. 1983. *The Culture of Time and Space 1880–1918.* Cambridge, MA: Harvard University Press.

Kropotkin, P. 1899. *Memoirs of a Revolutionist.* Boston, New York: The Riverside Press.

La Grye, Bouquet de, and C. Pujazon. 1897. "Différences de Longitudes entre San Fernando, Santa Cruz de Tenerife, Saint-Louis et Dakar." In *Annales du Bureau des Longitudes* 18.

La Porte, F. 1887. "Détermination de la longitude d'Haiphong (Tonkin) par le télégraphe" [29 August 1887]. In *Comptes rendus de l'Académie des Sciences* 105, pp. 404–6.

Lallemand, C. M. 1897. *L'unification internationale des heures et le système des fuseaux horaires*. Paris: Bureaux de la revue scientifique.

Lallemand, C. M. 1912. "Projet d'organisation d'un service international de l'heure." In *Annales du Bureau des Longitudes* 9, pp. D261–D268.

Landes, David S. 1983. *Revolution in Time. Clocks and the Making of the Modern World.* London; Cambridge, MA: Belknap Press (Harvard University Press).

Laue, M. 1913. *Das Relativitätsprinzip*. Braunschweig: Friedrich Vieweg & Sohn.

Le Livre du centenaire de la naissance de Henri Poincaré 1854–1954. Paris: Gauthier-Villars et Fils, 1955.

Loewy, M., F. Le Clerc, and O. de Bernardières. 1882. "Détermination des différences de longitude entre Paris–Berlin et entre Paris–Bonn." In *Annales du Bureau des Longitudes* 2.

Lorentz, H. A. 1904. "Electromagnetic Phenomena in a System Moving with Any Velocity Smaller Than That of Light." In *Proceedings of the Royal Academy of Amsterdam* 6, pp. 809–32.

Lorentz, H. A. 1937. "Versuch einer Theorie der electrischen und optischen Erscheinungen in bewegten Körpern." [1895]. In H. A. Lorentz, *Collected Papers*, Vol. 5. The Hague: Martinus Nijhoff, pp. 1–139.

Mach, Ernst. [1893]. *The Science of Mechanics: A Critical and Historical Account of Its Development*. Trans. Thomas J. McCormack. New Introduction by Karl Menger. La Salle: The Open Court Publishing Company, 1960.

Merle, U. 1989. "Tempo! Tempo! Die Industrialisierung der Zeit im 19. Jahrhundert." In *Uhrzeiten. Die Geschichte der Uhr und ihres Gebrauches*, ed. Igor A. Jenzen. Marburg: Jonas Verlag.

Messerli, Jakob. 1995. *Gleichmässig, pünktlich, schnell. Zeiteinteilung und Zeitgebrauch in der Schweiz im 19. Jahrhundert*. Zürich: Chronos Verlag.

Mestral, Ayman de. 1960. *Mathius Hipp 1813–1893, Jean-Jacques Kohler 1860–1930, Eugene Faillettaz 1873–1943, Jean Landry 1875–1940*. (Pionniers suisses de l'économie et de la technique 5.) Zürich: Boillat.

Mill, John S. 1965. *A System of Logic. Ratiocinative and Inductive. Being a Connected View of the Principles of Evidence and the Methods of Scientific Investigation*. London: Spottiswoode, Ballantyne & Co.

Miller, Arthur I. 1981. *Albert Einstein's Special Theory of Relativity. Emergence (1905) and Early Interpretation (1905–1911)*. London: Addison-Wesley Publishing Company, Inc.

Miller, Arthur I. 1982. "The Special Relativity Theory: Einstein's Response to the

Physics of 1905." In *Albert Einstein. Historical and Cultural Perspectives. The Centennial Symposium in Jerusalem*, eds. Gerald Holton and Yehuda Elkana. Princeton, NJ: Princeton University Press, pp. 3–26.

Miller, Arthur I. 1986. *Frontiers of Physics: 1900–1911. Selected Essays.* (With an Original Prologue and Postscript.) Basel: Birkhäuser.

Miller, Arthur I. 2001. *Einstein, Picasso. Space, Time, and the Beauty That Causes Havoc.* New York: Basic Books.

Moltke, Helmuth Graf von. 1892. "Dritte Berathung des Reichshaushaltsetats: Reichseisenbahnamt, Einheitszeit." In idem, *Gesammelte Schriften und Denkwürdigkeiten des General-Feldmarschalls Grafen Helmuth von Moltke,* Vol. VII: Reden. Berlin: Ernst Siegfried Mittler und Sohn, pp. 38–43.

Moltke, Helmuth Graf von. [1891–1893]. *Gesammelte Schriften und Denkwürdigkeiten des General-Feldmarschalls Grafen Helmuth von Moltke,* 8 Vols. Berlin: Ernst Siegfried Mittler und Sohn.

Myers, Greg. 1995. "From Discovery to Invention: The Writing and Rewriting of Two Patents." In *Social Studies of Science* 25, pp. 57–105.

Newton, Isaac. 1952. *Newton's Philosophy of Nature. Selections from His Writings.* New York: Hafner.

Nordling, M. W. de. 1888. "L'unification des heures." In *Revue générale des chemins de fer,* pp. 193–211.

"Le Nouvel étalon du mètre." In *Le Magasin Pittoresque* (1876), pp. 318–22.

Nye, Mary Jo. 1979. "The Boutroux Circle and Poincaré's Conventionalism." In *Journal of the History of Ideas* 40, pp. 107–20.

Olesko, Kathryn. 1991. *Physics as a Calling: Discipline and Practice in the Konigsberg Seminar for Physics.* Ithaca, NY: Cornell University Press.

Ozouf, Mona. 1992. "Calendrier." In *Dictionnaire critique,* eds. F. Furet and M. Ozouf. Paris: Flammarion, pp. 91–105.

Pais, Abraham. 1982. *"Subtle is the Lord . . ." The Science and the Life of Albert Einstein.* Oxford: Oxford University Press.

Paty, Michel. 1993. *Einstein Philosophe.* Paris: Presses Universitaires de France.

Pearson, Karl. 1892. *The Grammar of Science.* (With an Introduction by Andrew Pyle.) Bristol: Thoemmes Antiquarian Books, reprinted 1991.

Pestre, Dominique. 1992. *Physique et physiciens en France 1918–1940.* Paris: Gordon and Breach Science Publishers S.A.

Pickering, Edward. 1877. *Annual Report of the Director of Harvard College Observatory.* Cambridge: John Wilson & Son.

Picon, Antoine. 1994. *La Formation polytechnicienne 1794–1994.* Paris: Dunod.

Planck, Max. 1998. *Eight Lectures on Theoretical Physics.* Trans. A. P. Wills. Mineola, NY: Dover Publications.

Poincaré, Henri. 1881. "Mémoire sur les courbes définies par une équation différentielle (première partie)." In *Journal de Mathématiques pures et appliquées*, Ser. 3, 7, pp. 375–422.

Poincaré, Henri. [1885]. "Sur les courbes définies par les équations différentielles." In *Oeuvres*, Vol. 1 (1928), pp. 9–161.

Poincaré, Henri. [1887]. "Sur les hypothèses fondamentales de la géometrie." In *Oeuvres*, Vol. 11 (1956), pp. 79–91.

Poincaré, Henri. [1890]. "Sur le problème des trois corps." In *Oeuvres*, Vol. 7 (1952), pp. 262–490.

Poincaré, Henri. 1894. *Les oscillations électriques. Leçons professées pendant le premier trimestre 1892–1893.* Paris: Georges Carré.

Poincaré, Henri. [1897]. "La décimalisation de l'heure et de la circonférence." In *Oeuvres*, Vol. 8 (1952), pp. 676–79.

Poincaré, Henri. 1897. "Rapport sur les résolutions de la commission chargée de l'étude des projets de décimalisation du temps et de la circonférence." In *Commission de décimalisation du temps et de la circonférence*, pp. 1–12.

Poincaré, Henri. [1898] "La logique et l'intuition dans la science mathématique et dans l'enseignement." In *Oeuvres*, Vol. 11 (1956), pp. 129–33.

Poincaré, Henri. 1900. "Rapport sur le projet de revision de l'arc meridien de Quito" [25 July 1900]. In *Comptes rendus de l'Académie des Sciences*, CXXXI, pp. 215–36. Further reports on the Quito expedition (presented by Poincaré) follow in the same journal at yearly intervals through 1907.

Poincaré, Henri. [1900]. "La Dynamique de l'électron." In *Oeuvres*, Vol. 9 (1954), pp. 551–86.

Poincaré, Henri. [1900]. "La théorie de Lorentz et le principe de réaction." In *Oeuvres*, Vol. 9 (1954), pp. 464–93.

Poincaré, Henri. [1901]. *Électricité et optique. La lumière et les théories électrodynamiques.* Paris: Jacques Gabay, reprinted 1990.

Poincaré, Henri. 1902. "Notice sur la télégraphie sans fil." In *Oeuvres*, Vol. 10 (1954), pp. 604–22.

Poincaré, Henri. [1902]. *Science and Hypothesis.* With a Preface by J. Larmor. New York: Dover Publications, 1952.

Poincaré, Henri. 1903. "Le Banquet du 11 Mai." In *Bulletin de l'Association.* Paris: L'Université de Paris, pp. 57–64.

Poincaré, Henri. [1904]. "Etude de la propagation du courant en période variable sur une ligne munie de récepteur." In *Oeuvres*, Vol. 10 (1954), pp. 445–86.

Poincaré, Henri. 1904. "The Present State and Future of Mathematical Physics" (orig. "L'État actuel et l'avenir de la physique mathématique"). In *Bulletin des*

Sciences Mathématiques 28, pp. 302–24. (Partially reprinted in idem, *Valeur de la Science*. Paris: Flammarion, pp. 123–47.)

Poincaré, Henri. 1904. *Wissenschaft und Hypothese*. Deutsche Ausgabe mit erläuternden Anmerkungen von F. und L. Lindemann. Stuttgart: B.G. Teubner.

Poincaré, Henri. 1905. *La science et l'hypothèse*. Paris: Flammarion.

Poincaré, Henri. 1909. "La mécanique nouvelle." In *La Mécanique nouvelle*. Paris: Jaques Gabay, 1989, pp. 1–17.

Poincaré, Henri. 1910. "Cornu." In idem, *Savants et Écrivains*, pp. 103–24.

Poincaré, Henri. 1910. "La mécanique nouvelle." In *Sechs Vorträge über ausgewählte Gegenstände aus der reinen Mathematik und mathematischen Physik*. Leipzig, Berlin: B.G. Teubner, pp. 51–58.

Poincaré, Henri. 1910. "Les Polytechniciens." In idem, *Savants et Écrivains*, pp. 265–79.

Poincaré, Henri. 1910. *Savants et Écrivains*. Paris: Flammarion.

Poincaré, Henri. 1912. "General Conclusions." In *La théorie du rayonnement et les quanta. Rapports et discussions de la réunion tenue à Bruxelles, du 30 octobre au 3 novembre 1911*, eds. P. Langevin and M. de Broglie. Paris: Gauthier-Villars, pp. 451–54.

Poincaré, Henri. 1913. "Mathematical Creation." In idem, *The Foundations of Science*, pp.383–94.

Poincaré, Henri. 1913 "The Measure of Time." In idem, *The Foundations of Science*, pp. 223–34.

Poincaré, Henri. [1913]. "The Moral Alliance." In idem, *Mathematics and Science: Last Essays*. Trans. J. W. Bolduc. New York: Dover Publications, 1963.

Poincaré, Henri. *Oeuvres*, Vol. 1 (1928)–Vol. 11 (1956). Paris: Gauthier-Villars.

Poincaré, Henri. 1952. "Rapport sur la proposition d'unification des jours astronomique et civil." In *Oeuvres*, Vol. 8, pp. 642–47.

Poincaré, Henri. 1953–54. "Les Limites de la loi de Newton." In *Bulletin Astronomique* 17, pp. 121–78, 181–269.

Poincaré, Henri. [1913]. "Space and Time." In idem, *Mathematics and Science: Last Essays (Dernières pensées)*. Trans. J. W. Bolduc. New York: Dover Publications, 1963, pp. 15–24.

Poincaré, Henri. 1970. "La mesure du temps." In idem, *La valeur de la science*. Préface de Jules Vuillemin. Paris: Flammarion.

Poincaré, Henri. 1982. *The Foundations of Science* [1913]. Authorized trans. George Bruce Halsted (with a special Preface by Poincaré and an Introduction by Josiah Royce). Washington: University Press of America, Inc.

Poincaré, Henri. 1993. *New Methods of Celestial Mechanics*. (History of Modern

Physics and Astronomy 13). Ed. Daniel L. Goroff. Boston: American Institute of Physics.

Poincaré, Henri. 1997. *Trois suppléments sur la découverte des fonctions fuchsiennes. Three Supplementary Essays on the Discovery of Fuchsian Functions. Une édition critique des manuscrits avec une introduction. A Critical Edition of the Original Manuscripts with an Introductory Essay.* Ed. Jeremy J. Gray and Scott A. Walter. Berlin, Paris: Akademie-Verlag Berlin/Albert Blachard.

Poincaré, Henri. 1999. *La Correspondance entre Henri Poincaré et Gösta Mittag-Leffler. Avec en annexes les lettres échangées par Poincaré avec Fredholm, Gyldén et Phragmén.* Présentée et annotée par Philippe Nabonnand. Basel: Birkhäuser Verlag.

Poincaré, Henri, Jean Darboux, and Paul Appell. 1908. *Affaire Dreyfus. La Revision du Procès de Rennes. Enquête de la Chambre Criminelle de la Cour de Cassation,* Vol. 3. Paris: Ligue Française pour la défense des droits de l'homme et du citoyen, pp. 499–600.

Prescott, George B. [1866]. *History, Theory, and Practice of the Electric Telegraph.* Cambridge: Cambridge University Press, reprinted 1972.

Proceedings of the General Time Convention, Chicago, October 11, 1883. New York: National Railway Publication Company, 1883.

Proceedings of the Southern Railway Time Convention, New York, October 17, 1883.

Pyenson, Lewis. 1985. *The Young Einstein. The Advent of Relativity.* Bristol, Boston: Adam Hilger.

Quine, Willard Van Orman. 1990. *Dear Carnap, Dear Van: The Quine-Carnap Correspondence and Related Work.* Ed. Richard Creath. Berkeley: University of California Press.

Rayet, G., and Lieutenant Salats. 1890. "Détermination de la longitude de l'observatoire de Bordeaux." In *Annales du Bureau des Longitudes* 4.

Renn, Jürgen, and Robert Schulmann (eds.). 1992. *Albert Einstein–Mileva Marić. The Love Letters.* Trans. Shawn Smith. Princeton, NJ: Princeton University Press.

Renn, Jürgen. 1997. "Einstein's Controversy with Drude and the Origin of Statistical Mechanics: A New Glimpse from the 'Love Letters'." In *Archive for History of Exact Sciences* 51, pp. 315–54.

"Report to the Board of Visitors, Nov. 4, 1864." In *Astronomical Observations Made at the Royal Observatory in Edinburgh* 13 (1871), pp. R12–R20.

"Report of the Committee on Standard Time" [May 1882–Dec. 1882]. In *Proceedings of the American Metrological Society* 3 (1883), pp. 27–30.

"Report of the Committee on Standard Time, May 1879" [Dec. 1878–Dec. 1879]. In *Proceedings of the American Metrological Society* 2 (1882), pp. 17–44.

"Report of the Director to the Visiting Committee of the Observatory of Harvard University." In *Annals of Astronomical Observatory of Harvard College*, Vol. 1.

Report of the Superintendent of the Coast Survey, Showing the Progress of the Survey During the Year 1860 (resp. 1861, 63, 64, 65, 67, 70, 74). Washington: U.S. Government Printing Office, 1861 (resp. 1862, 64, 66, 67, 69, 73, 77).

Rothé, Edmond. 1913. *Les applications de la Télégraphie sans fil: Traité pratique pour la réception des signaux horaires*. Paris: Berger-Levrault.

Roussel, Joseph. 1922. *Le premier livre de l'amateur de T.S.F.* Paris: Vuibert.

Roy, Maurice, and René Dugas. 1954. "Henri Poincaré, Ingénieur des Mines." In *Annales des Mines* 193, pp. 8–23.

Rynasiewicz, Robert. 1995. "By Their Properties, Causes and Effects: Newton's Scholium on Time, Space, Place and Motion, Part I: The Text." In *Studies in History and Philosophy of Science* 26, pp. 133–53; "Part II: The Context," pp. 295–321.

Sarrauton, Henri de. 1897. *L'heure décimale et la division de la circonférence*. Paris: E. Bernard.

Schaffer, Simon. 1992. "Late Victorian Metrology and Its Instrumentation: A Manufactory of Ohms." In *Invisible Connections. Instruments, Institutions, and Science*, eds. Robert Bud and Susan E. Cozzens. Washington: Spie Optical Engineering Press, pp. 23–56.

Schaffer, Simon. 1997. "Metrology, Metrication and Victorian Values." In *Victorian Science in Context*, ed. Bernard Lightman. Chicago: The University of Chicago Press, pp. 438–74.

Schanze, Oscar. 1903. *Das schweizerische Patentrecht und die zwischen dem Deutschen Reiche und der Schweiz geltenden patentrechtlichen Sonderbestimmungen*. Leipzig: Harry Buschmann.

Schiavon, Martina. n.d. "Savants officiers du Dépôt général de la Guerre (puis Service Géographique de l'Armée). Deux missions scientifiques de mesure d'arc de méridien de Quito (1901–1906)." In *Revue Scientifique et Technique de la Défense*, forthcoming.

Schilpp, Arthur Paul (ed.). 1963. *The Philosophy of Rudolf Carnap*. (The Library of Living Philosophers, Vol. XI.) London: Cambridge University Press.

Schilpp, Arthur Paul. 1970. *Albert Einstein: Philosopher-Scientist*, two Vols. La Salle: Open Court.

Schlick, Moritz. 1987. "Meaning and Verification." In idem, *Problems of Philosophy*. (Vienna Circle Collection 18), ch. 14, pp. 127–33.

Schlick, Moritz. 1987. *The Problems of Philosophy in Their Interconnection. Winter Semester Lectures, 1933–34*. Eds. Henk L. Mulder, A. J. Kox, and Rainer Hegselmann. Boston: D. Reidel Publishing Company.

Schmidgen, Henning. n.d. *Time and Noise: On the Stable Surroundings of Reaction Experiments (1860–1890)*, forthcoming.

Seelig, Carl (ed.). 1956. *Helle Zeit—Dunkle Zeit. Jugend-Freundschaft-Welt der Atome. In Memoriam Albert Einstein.* Zürich: Europa Verlag.

Septième conférence géodésique internationale. Rome: Imprimerie Royale D. Ripamonti, 1883.

Shaw, Robert B. 1978. *A History of Railroad Accidents. Safety, Precautions, and Operating Practices.* Binghamton, NY: Vail-Ballou Press.

Sherman, Stuart. 1996. *Telling Time. Clocks, Diaries, and English Diurnal Form, 1660–1785.* London, Chicago: The University of Chicago Press.

Shinn, Terry. 1980. *Savoir scientifique et pouvoir social. L'École Polytechnique.* Préface de François Furet. Paris: Presses de la Fondation Nationale des Sciences Politiques.

Shinn, Terry. 1989. "Progress and Paradoxes in French Science and Technology 1900–1930." In *Social Science Information* 28, pp. 659–83.

Smith, Crosbie, and M. Norton Wise. 1989. *Energy and Empire: A Biographical Study of Lord Kelvin.* Cambridge: Cambridge University Press.

Sobel, Dava. 1995. *Longitude. The True Story of a Lone Genius Who Solved the Greatest Scientific Problem of His Time.* New York: Walker and Company.

Staley, Richard. 2002. "Travelling Light." In *Instruments, Travel and Science*, eds. Marie-Noëlle Bourguet, Christian Licoppe, and H. Otto Sibum. New York: Routledge.

Stephens, Carlene E. 1987. "Partners in Time: William Bond & Son of Boston and the Harvard College Observatory." In *Harvard Library Bulletin* 35, pp. 351–84.

Stephens, Carlene E. 1989. "'The Most Reliable Time': William Bond, the New England Railroads, and Time Awareness in the 19th-Century America." In *Technology and Culture* 30, pp. 1–24.

Taylor, Edwin, and John Wheeler. 1966. *Spacetime Physics.* New York: W.H. Freeman.

Urner, Klaus. 1980. "Vom Polytechnikum zur Eidgenössischen Technischen Hochschule: Die ersten hundert Jahre 1855–1955 im Ueberblick." In *Eidgenössische Technische Hochschule Zürich. Festschrift zum 125jährigen Bestehen (1955–1980).* Zürich: Verlag Neue Zürcher Zeitung, pp. 17–59.

Walter, Scott. 1999. "The non-Euclidean Style of Minkowskian Relativity." In *The Symbolic Universe*, ed. Jeremy Gray. Oxford: Oxford University Press.

Warwick, Andrew. 1991/1992. "On the Role of the FitzGerald-Lorentz Contraction Hypothesis in the Development of Joseph Larmor's Electronic Theory of Matter." In *Archive for History of Exact Sciences* 43, pp. 29–91.

Warwick, Andrew. 1992/1993. "Cambridge Mathematics and Cavendish Physics: Cunningham, Campbell and Einstein's Relativity 1905–1911. Part I: The Uses of Theory." In *Studies in History and Philosophy of Science* 23, pp. 625–56; "Part II: Comparing Traditions in Cambridge Physics." In idem, 24 (1993), pp. 1–25.

Weber, R., and L. Favre. 1897. "Matthäus Hipp, 1813–1893." In *Bulletin de la société des sciences naturelles de Neuchâtel* 24, pp. 1–30.

Welch, Kenneth F. 1972. *Time Measurement. An Introductory History*. Baskerville: Redwood Press Limited Trowbridge Wiltshire.

Whittaker, Edmund. 1953. *A History of the Theories of Aether and Electricity. Vol. II: The Modern Theories 1900–1926*. New York: Harper & Brothers, reprinted 1987.

Wise, Norton M. 1988. "Mediating Machines." In *Science in Context* 2, pp. 77–113.

Wise, Norton M. (ed.). 1995. *The Values of Precision*. Princeton, NJ: Princeton University Press.

Wise, Norton M., and David C. Brock. 1998. "The Culture of Quantum Chaos." In *Studies in the History and Philosophy of Modern Physics* 29, pp. 369–89.

INDEX

Page numbers in *italics* refer to illustrations.